Nanotechnology
Intellectual Property Rights

Research, Design, and Commercialization

PERSPECTIVES IN NANOTECHNOLOGY

Series Editor
Gabor L. Hornyak

Nanotechnology
Intellectual Property Rights

Research, Design, and Commercialization

Prabuddha Ganguli
Siddharth Jabade

CRC Press
Taylor & Francis Group
Boca Raton London New York

CRC Press is an imprint of the
Taylor & Francis Group, an **informa** business

CRC Press
Taylor & Francis Group
6000 Broken Sound Parkway NW, Suite 300
Boca Raton, FL 33487-2742

© 2012 by Taylor & Francis Group, LLC
CRC Press is an imprint of Taylor & Francis Group, an Informa business

No claim to original U.S. Government works

Visit the Taylor & Francis Web site at
http://www.taylorandfrancis.com

and the CRC Press Web site at
http://www.crcpress.com

Contents

List of Illustrations

List of Tables

Foreword

Any rapidly developing cutting-edge technology carries along with it extensive intellectual property issues. This is particularly true for nanotechnology, which is all-pervasive with its reach ranging from materials and electronics to biotechnology and health care. Any new technology also can cause unexpected and of unwarranted concerns, as was the case with nanobiotechnology when the novel *Prey* by Michael Crichton was published.

The frenetic pace of research and development of this field, combined with an early recognition of its potential applications and commercialization, has led to widespread start-ups and has attracted funding from governments, angel investors, venture capital agencies, financial institutions and industries.

Being extensively knowledge-led, the competitive advantage in nanotechnology is strongly correlated with gaining and maintaining exclusivity of one's acquired technology knowledge in the marketplace. It is in this context that intellectual property rights and especially patents become significant, with their role in knowledge sharing, knowledge protection and enforcement, and accrual of the benefits of commercialization for business sustenance.

The book title, *Nanotechnology and Intellectual Property Rights: Research, Design, and Commercialization* by Prabuddha Ganguli and Siddharth Jabade comprehensively, and yet exhaustively, captures the essence of the dynamics of the development and commercialization in this field, where strategic management of IPR has become imperative. A unique feature of the book is its cogent articulation of techno-legal aspects of innovations.

This book, in six chapters, illustrates how patents have been used in this sunrise sector by technology developers, business entrepreneurs, funding agencies, and industries for seeding, nurturing, and sustaining nanotechnology-enabled businesses. The chapter on patent-led nanotechnology businesses lucidly demonstrates the significance of national and institutional IPR policies and their implementation in fructifying academic–industry collaborations, managing intraindustry and interindustry partnerships, and IP transactions, in both the precompetitive technology development and commercialization phases. The chapter entitled Interfacing with Nanofuture tries to landscape the nanoworld of the "day after," linking it with evolutionary dynamics, the function of asset value of patent portfolios in enterprise restructuring, lessons learned thus far, the gray areas in patent-related issues that need to be address, and all that would be required by learning enterprises to adapt, adopt, plan, and implement to meet future nanochallenges.

I compliment Prabuddha Ganguli and his co-author Siddharth Jabade for this invaluable and creative endeavor, which will fill the present void of an authoritative reference source in nanotechnology-related IPR. I am sure it will be used extensively as well by students, research workers, entrepreneurs, business management specialists, lawyers, and policy makers involved in the field of nanotechnology.

R. Chidambaram
Principal Scientific Adviser
Government of India

Preface

Nucleation and the growth of businesses in nanotechnology have created an intensely complex intellectual property (IP) landscape with competing players vying for exclusivity in an already overcrowded IP space. Nanotechnology provides a unique opportunity to tune and retune technology development and transfer IP management including business strategies at every stage of the knowledge value chain from concepts to commercialization.

More often than not, high technology–led fields operate in disjointed and demarcated silos of R&D and businesses without systematic dynamic linking with organized intellectual property management systems, thereby suffering from the risk of weak IP protection of developed technologies, blurred freedom to operate in markets of business interests, unforeseen IP litigations, improper asset valuations, etc. Ensuring success of "nanobusinesses" in a fiercely competitive world of "the day after" will require skill, ability, and tact to conceptualize applications, realize them cost effectively, and even interactively cooperate with competing stakeholders when necessary in a multifarious IP and regulatory ecosystem. Researching this book has been an exciting experience as we had to fathom the depths of nanotechnology to unravel the intrinsic connectivities.

This book titled *Nanotechnology and Intellectual Property Rights: Research, Design, and Commercialization* has been designed to comprehensively address interrelated issues in an integrated and comprehensive manner.

The first three chapters illustrate the evolving patent landscape in nanotechnology, patenting procedures as applied in diverse jurisdictions, searching for nanotechnology prior art including the creation of search strategies, and the international patent classification system. The fourth chapter on patent-led nanotechnology businesses provides a spectrum of perspective learning using a plethora of case studies involving the building of valuable patent portfolios, growth of start-ups, consolidation of IP-led nanobusinesses through mergers, acquisitions, joint ventures, strategic investments, etc. The fifth chapter deals with patent litigations in nanotechnologies and exemplifies the significance of the strategic crafting of agreements related to IP transactions, compliance of contractual obligations, importance of well-drafted patent specifications, sensitivities in conducting techno-legal due diligence prior to the development and marketing of products, vulnerabilities in challenging/defending the validity of patents and negotiating settlements. Chapter 6 gazes into the future of IP landscaping in nanotechnology especially in terms of the gray areas on patentability, public perceptions of risks to health and ecosystems, institutionalized management of intellectual property rights, and the steps that will be necessary to meet such challenges.

Capacity building in IP will continue to be a key factor that will determine the success of businesses and the model IPRinternalise® has been described for institutional adoption to continually build human and infrastructural resources to meet the emerging and unforeseen challenges of "the day after."

Writing this book has been a wonderful and insightful learning experience for us. It is our fond belief that readers across all spheres of activities linked with nanotechnology will find this endeavor a catalytic facilitator to seamlessly integrate IPR and especially patents to make their nanodreams turn into profitable realities.

Acknowledgments

This book was conceptualized during a discussion between Prabuddha Ganguli, Siddharth Jabade, Professor Louis Hornyak, and Professor Joydeep Dutta on ways and means to seamlessly integrate IPR into R&D and commercialization processes, especially in a high-technology area such as nanotechnology. Though several books and resource materials have dealt with this subject, the need for a comprehensive volume linking the key features in the knowledge value chain was strongly felt. We are indebted to Professor Hornyak and Professor Dutta of the Asian Institute of Technology (Bangkok) for their invaluable contributions to the creation of this book.

Dominic Keating, director of the Intellectual Property Attaché Program, United States Patent and Trademark Office provided us with the USPTO patenting data in Class 977. Denis Dambois, DG Trade of the European Commission furnished us with the links to the *Observatory NANO Factsheets.* Michael Blackman, editor of the journal *World Patent Information (WPI)* enriched our research by giving us access to several relevant articles in nanotechnology and patents. Makarand Waikar and Dr. Jyoti Singh of Sci-Edge Information, SciFinder/STN/Chemical Abstract Service (CAS) representative in India, augmented the contents of the chapter on prior art search with their expertise. We express our heartfelt thanks to all of them for their prompt response to our requests.

We are grateful to Dr. R. Chidambaram, principal scientific adviser to the government of India for his in-depth analysis and thought-provoking foreword to this book.

We have benefited immensely from the contributions of various authors on the World Wide Web. We express our heartfelt thanks to them.

It has been a gratifying experience working with Jill Jurgensen, Nora Konopka and the entire team of the CRC Press. Their constant professional support has led to the speedy publishing of this book.

Our families gave much needed encouragement at all times during the course of this project and no words of appreciation are adequate to express our gratitude to them. Prabuddha's wife Subhra and daughter Rinki also provided their invaluable eagle eye, helping to bring lucidity to every stage of this writing. Siddharth thanks his wife Vaishali and son Shreyan for their understanding and encouraging smiles that continue to fuel his enthusiasm.

Prabuddha Ganguli and Siddharth Jabade

About the Authors

Prabuddha Ganguli, PhD, is the CEO of VISION-IPR offering services in management of intellectual property rights, information security and knowledge management including designing of innovation processes, strategizing technology transfers, and conducting IP due diligence for joint ventures, mergers, and acquisitions. He is a qualified patent agent in India.

He is an Honorary Scientific Consultant for Innovation and IPR matters to the Office of the Principal Scientific Adviser, Government of India; a member of the advisory board of the International Intellectual Property Institute, Washington; and a member of the advisory board of the Institute of Intellectual Property Studies, Mumbai. He is a member of the international editorial board of the IPR journal *World Patent Information*.

Dr. Ganguli has over 50 publications in technical fields, over 60 publications in IPR, and 5 books, namely *Gearing up for Patents—The Indian Scenario*, Universities Press (1998); *Intellectual Property Rights—Unleashing the Knowledge Economy*, Tata McGraw-Hill (2001); *Shaping the Future*, UNIDO-WIPO-IMTMA (2005); *Technology Transfer Issues in Biotechnology—A Global Perspective*, co-edited with Dr. Ben Prickril and Dr. Rita Khanna, published by Wiley-VCH (Germany) (2009); and *Geographical Indications—Its Evolving Contours*, Institute of Intellectual Studies (2009).

As an international consultant to the World Intellectual Property Organization (WIPO), Geneva, Dr. Ganguli is actively involved in IPR capacity–building programs, and formulation and implementation of national IPR policies in several developing and least developed countries.

After obtaining a PhD in chemistry from the Tata Institute of Fundamental Research, Mumbai, he did postdoctoral research in Germany as an Alexander von Humboldt Foundation Fellow and in Canada as a research associate. Following a brief stint as visiting scientist at the Bhabha Atomic Research Centre, Government of India, he worked in Hindustan Unilever Ltd., the Indian subsidiary of Unilever for over two decades in various management positions including research, manufacturing, IPR, knowledge management, corporate information security, and corporate planning. He started his consulting firm VISION-IPR in 2001.

In February 2011, he was awarded the Chemtech Pharma-Bio World Award 2011 for outstanding contribution in the field of intellectual property.

Siddharth Jabade, PhD, possesses a unique blend of technical and patent expertise. He is presently professor of mechanical engineering at the Vishwakarma Institute of Technology, Pune, India (VI, Pune). He completed his bachelor's and master's degrees in engineering at the University of Pune and obtained his PhD from the Indian Institute of Technology, Bombay (IIT Bombay). He has been a senior associate at VISION-IPR, a leading intellectual property management firm in India. He is the author of several scientific publications and  inventor in patents related to inventions in heat transfer. He is also the coordinator, Intellectual Property Rights Facilitation Centre at VI, Pune. He is a qualified patent agent in India and is involved in drafting and prosecuting patent applications.

Dr. Jabade has lectured as a resource person in Train the Trainers IPR program in India conducted by the World Intellectual Property Organization (WIPO). He has provided support as a technical expert in patent infringement cases and drafted numerous patent applications in diverse technologies. He has been a key resource person in IPR capability development programs and has been a regular visiting faculty at the Asian Institute of Technology, Bangkok and at the Institute of Intellectual Property Studies (IIPS), Mumbai. Dr. Jabade has contributed significantly to the building of new platforms for higher education and their implementation, such as IPRinternalise®, in collaboration with Dr. Prabuddha Ganguli to seamlessly merge intellectual property rights in educational systems. His lead role in setting up the Innovation Management Process at the Center of Excellence in Nanotechnology (CoEN) at Asian Institute of Technology, Bangkok, has set the path to early identification and management of inventions for commercial success through the establishment of a sustainable network involving diverse stakeholders/funding agencies. Dr. Jabade is widely traveled as a researcher and is a passionate teacher of engineering, technology transfer, and intellectual property rights.

1

How Big Is Small?

The innate human urge to create a better tomorrow has driven the world to continually innovate and compete. The last few decades have experienced the cross-fertilization of disciplines with the emergence of convergent technologies, which in due course have hatched and grown into novel knowledge-led businesses, one of them being nanotechnology.

The contours of the nanotechnology domain have been changing at amoebic frequency ever since its beginning. The science of "nano" (generally as small as 0.1 to 100 nm; 1 nm = 10^{-9} m) has already permeated all possible fields of applications with ideations inching their way into innovations for possible realization in the marketplace.

In 1959 in his talk titled, "There's Plenty of Room at the Bottom," Professor Richard Feynman suggested the concept that "ultimately—in the great future—we can arrange the atoms the way we want; the very atoms, all the way down!" Such a thought of tailoring materials at will from the very basic building blocks was further christened as "nanotechnology" by Professor Norio Taniguchi in 1974: "'Nano-technology' mainly consists of the processing, separation, consolidation, and deformation of materials by one atom or one molecule."[1]

The inclusive nature of nanotechnology gives it a very special status as it mothers innovations to deliver inventions to provide a host of new, pure, and hybrid options by way of nanobiotechnology, nanostructures, nanocomposites, nanomedicine, nanotaggants for security systems, nanoelectronics, nanodevices, etc. This nano-inclusiveness creates the ecosystem to incite and integrate human talents and energy for global participation at all levels.

Such progressive developments are and will only be possible through dynamic global enterprise networking (Figure 1.1) of emerging and created markets, industry, academic institutions/universities, R&D organizations, and governments.

Academic institutions are wellsprings of knowledge with the goal of dynamically stimulating, fostering, and incubating minds. For successful enterprise and participative enterprise networking, one has to consciously guard against the dangers of academic institutions coalescing into "black holes of knowledge" with the inadvertent creation of virtual walls inhibiting meaningful knowledge transaction across academic boundaries into the demanding operational space.

For the sustenance of an innovative culture, academic institutions on one hand have to be involved in uninterrupted creation, enrichment, and

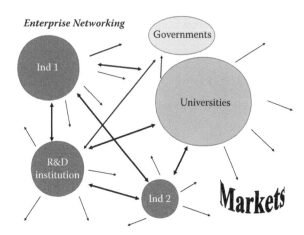

FIGURE 1.1
Enterprise knowledge networking.

maintenance of contextual educational programs with sustained knowledge renewal for the development of well-prepared human resources, and on the other hand participate in maximizing speedy transformation of knowledge into tangibles for industries to execute commercial exploitation.

The industry sector has to link into the value-added knowledge networking as key irrigating channels for effective nurturing and productive harvesting of the knowledge-deliverables to the markets for societal benefit.

The process of "knowledge generation" and transfer demands innovative frameworks to justify ownership of the developed knowledge and benefit sharing between contributing partners, thereby providing pathways to incubation of minds to markets.

Governments are required to play a facilitating role in this dynamic enterprise, networking through enabling policies for the key actors to perform in an orchestrated manner to make the benefits of innovation finally available to the public.

De-cocooned enterprise networking demands tuning and retuning of intra- and interorganizational teamwork to craft and capitalize on opportunities without "reinventing the wheel," with assured freedom to operate with minimized risk. Designing enabling conduits for innovation flow with the option to pole vault innovations from their centers of creation to markets now requires the strategic management of Intellectual Property Rights (IPR) portfolios seamlessly built into the innovation process.

The promising potential of this field has set the governments around the world to continually invest approximately $10 billion per year with annual growth of 20% in the development of this field in all its diversity with the expectation to reap effective economic benefits for society in the near future. Joining the United States and European Commission Initiatives in Nanotechnology, the governments in the Asia Pacific Region have set up significant investments in nanotechnology since 2001. The Russian government also has taken up nano-technology since 2007, which is being managed by the Russian Corporation of Nanotechnologies (RUSNANO).[2] By the end of 2011, the total government funding for nanotechnology research worldwide was $65 billion, and expected to rise to $100 billion by 2014. Figure 1.2 shows the investment profile by governments in the field of nanotechnology.

The dominating investments are by the United States followed by Japan and the European Union (EU). Within the EU, Germany, France, and the United Kingdom lead public investments. China, Taipei, and South Korea have invested significantly in nanotechnology research over the past five years. The ratio of private to public funding for nanotechnology research in Japan, the United States, and the EU is 1.7, 1.2, and 0.5, respectively. By 2015, private and government funding is expected to be approximately $250 billion (Figure 1.2).

World Intellectual Property Report 2011,[3] released in November 2011, states that the innovation process is increasingly international in nature and emphasizes that "innovation is seen to have become more collaborative and open…". The report further states,

> In particular, firms practicing open innovation strategically manage inflows and outflows of knowledge to accelerate internal innovation and to expand the markets for external uses of their intangible assets. "Horizontal" collaboration with similar firms is one important element of open innovation, but it also includes "vertical" cooperation with customers, suppliers, universities, research institutes and others.

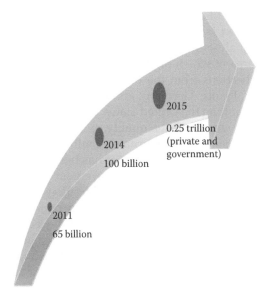

FIGURE 1.2
Investment profile by governments in nanotechnology.

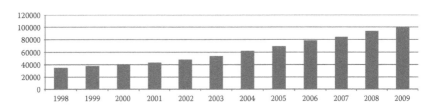

FIGURE 1.3
Total worldwide nanotechnology publications (Observatory NANO Factsheets, March 2011).

Knowledge sharing and peer group recognition is a key incentive among knowledge seekers and research publications in all forms continue to be the lead venting mechanism. Since the conceptualization of "nano," the scientific community has been on the Formula One track "Nanoathon" racing ahead with new findings, which is evident from the number of publications involving the science of nano depicted in Figure 1.3.

An Organization of Economic Cooperation and Development (OECD) report[4] indicates 63% of the global nanotechnology-related publications are contributed by the United States, Japan, Germany, France, United Kingdom, and China. Interestingly, Japan, China, Germany, France, Korea, Russian Federation, India, Taipei, Poland, Brazil, and Singapore have higher shares of nanotechnology publications relative to that of all publications from those countries, showing that nanotechnology has become a major thrust area for R&D. In contrast, countries such as the United States, United Kingdom, Canada, and Australia have higher shares of all publications relative to that of nanotechnology publications, even though nanotechnology is a major thrust area supported by the respective governments. The OECD report also presents an analysis in terms of the highly cited publications in the nanotechnology area in which the United States significantly stands out as compared to publications from the other countries. Performance of countries such as Japan, France, Germany, and the United Kingdom follows, but is substantially lower than the United States. However, China is fast catching up with the leading nations in the quality of its publications and citations in science and technology (S&T) journals. Nonlinear evolution of the frontiers of nanoscience and nanotechnologies has been due to the blurring of disciplinary boundaries and creation of convergent technologies as is well illustrated in the contents of the scientific publications (Figure 1.4) in which the fields that converge are typically material science, metallurgy, chemistry, physics of condensed matter, polymer science, electrical engineering, electronics, instrumentation, and biology. New fields, such as plasmonics, metamaterials, spintronics, graphene, cancer detection and treatment, drug delivery, synthetic biology, neuromorphic engineering, and quantum information systems, also have sprouted up in recent times. The last decade has witnessed as well the growth of international S&T networks and intergovernmental cooperative programs in nanotechnologies.

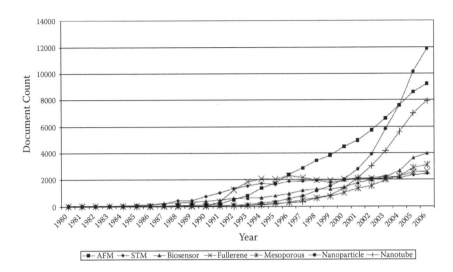

FIGURE 1.4
Number of publications in selected fields of nanotechnology. (From Finardi, U., Time-space analysis of scientific citations in patents: science-innovation links in the nano-technological paradigm, paper presented at the 6th Annual Conference of the EPIP Association Fine-Tuning IPR Debates, September 2011, http://www.epip.eu/conferences/epip06/papers/Parallel%20 Session%20Papers/FINARDI%20Ugo.pdf)

Concurrent with the scientific discoveries in nanotechnology (generative phase) has been the application of the newly understood principles to develop novel products and processes with diverse applications (application phase), creating proprietary knowledge domains and strategically trading them (trading phase). The challenge before industry especially in the trading phase continues to be in arriving at cost-effective solutions to make affordable products and processes available in the marketplace. In the case of selected innovations in nanotechnology ("nanovations"), the time gap between the generative, application, and trading phases has significantly collapsed making management of knowledge, including intellectual property rights, an absolute imperative (Figure 1.5).

Protection of intellectual property is of particular importance to the nanotechnology industry because of its complex knowledge matrix. To differentiate the products and services of the competitive stakeholders in different jurisdictions, strategic management of intellectual property rights through a combination of patent, industrial design registrations, trademark, copyright, and trade secret, as well as employee and third-party nondisclosure and assignment agreements is significant in a dense competitive landscape.

$$Competition \sim f[\,(knowledge).(creativity).(vision).(action)\,].ipr$$

$$\sim f[innovation].ipr$$

FIGURE 1.5
Dynamic nanovation.

Innovations per se do not automatically enjoy statutory protection. Among other factors that contribute to innovation, the product of knowledge, creativity, vision, and action plays a dominant role. Innovations, when operated on by the tools of IPR, provide the applicant statutory protection to stop others from exploiting that protected innovation without authorization of the owner of the right. Failure to obtain or maintain adequate protection of one's innovations through appropriate intellectual property rights for any reason, including through their prosecution process or in the event of litigation related to such intellectual property may jeopardize the interests of the innovator and the owner of the IPR. Further inappropriate management of IPR would only add to the costs of obtaining or protecting intellectual property and would then negatively impact operating results. Therefore, integration of IPR in the knowledge generation, transfer, and commercialization process needs to be mastered and executed for successful running of enterprises.

A recent study by Finardi[5] using an aggregate of citations of scientific journal articles in patents related to several nanotech items, the chronological distance between citing and cited, and the endogeneity of cited literature was performed. The study concluded that nanotech items show a common behavior, with a (relatively) short time lag of three to four years between publication of scientific articles (and thus codification of scientific knowledge) and citations into patent (and thus being incorporated into a technical application). In such a high knowledge-intensive paradigm, endogenous patents are cited more than endogenous scientific literature.

Nanotechnology in the marketplace has already started to impact every sphere of our lives as "nanovations" are embedded in various consumer products. As of March 2011, the nanotechnology consumer products inventory contains 1317 products or product lines.[6] Since the inception of the database in 2005 with 54 consumer products, the inventory has grown year by year with 2006 (356), 2007 (380), 2008 (803), 2009 (1015), and 2010 (1317) showing a growth in the inventory by 521%. The numbers in parentheses indicate the number of products. These products span diverse product categories, such as health and fitness (738), home and garden (209), automobile (126), food and beverages (105), multifunctional cross cutting (82), electronics and computers (59), appliances (44), and goods for children (30). The subcategories associated with the largest main category, health and fitness, include cosmetics (143 products), clothing (182), personal care (267), sporting goods (119), sunscreen (33), and filtration (43). Again, products with relevance to multiple categories have been accounted for multiple times. The cosmetics, clothing, and personal care subcategories are now the largest in the inventory. The breakdown of products by region indicates that companies based in the United States have the most products, with a total of 587, followed by companies in Europe (United Kingdom, France, Germany, Finland, Switzerland, Italy, Sweden, Denmark, the Netherlands) (367), East Asia (including China, Taiwan, Korea, and Japan) (261), and elsewhere around the world (Australia, Canada, Mexico, Israel, New Zealand, Thailand, Singapore, the Philippines, and Malaysia) (73). Two products have no country designation. There is a small set of materials explicitly referenced in nanotechnology consumer products. The most common material mentioned in the product descriptions is now silver (313 products). Carbon, which includes fullerenes, is the second most referenced (91), followed by titanium (including titanium dioxide) (59), silica (43), zinc (including zinc oxide) (31), and gold (28).

Analysis of the literature to date clearly outlines the application areas of nanoscience and nanotechnology for commercialization spread over aerospace/transportation, agrifoods, chemistry and materials, construction, energy, environment, health and medicine, security, IT and telecommunications, and textiles (Figure 1.6).

Recent estimates from diverse sources for the commercialization of the "nanogalaxy" of products incorporating nanotechnology until 2015 is expected to reach $2.41 trillion, as illustrated in Figure 1.7.

The global market value for nanotechnology is estimated to be $27 billion in 2015, for a five-year compound annual growth rate (CAGR) of 11.1%. The largest segment of the market, made up of nanomaterials, is expected to increase at a five-year CAGR of 14.7%, from nearly $10 billion in 2010 to nearly $19.6 billion in 2015. The second-largest segment, nanotools, is expected to reach a value of more than $6.8 billion in 2015. Thus, the five-year CAGR is projected to be 3.3%. The smallest segment of the market, nanodevices, will have the highest five-year CAGR, at 45.9%. It is projected to increase from an estimated $35.4 million in 2010 to nearly $234 million in 2015. The market for quantum dots (QDs), which in 2010 was estimated to generate $67 million in revenues, has been projected to grow over the next five years at a CAGR of 58.3%, to about $670 million by 2015. Nano-enabled batteries are expected to reach $2.5 billion by 2015. Nanostructured coatings and thin films in 2011 were forecasted to be approximately $227.5 million and expected to reach approximately $1.4 billion by 2016. The carbon nanotubes market in Western Europe is projected to reach $43.1 million by 2012. Nanocomposites has

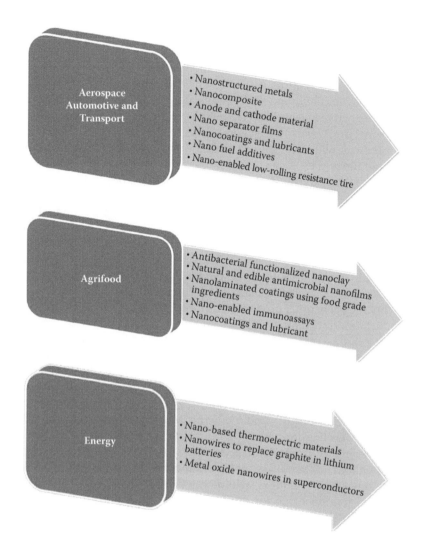

FIGURE 1.6
Emerging opportunities in nanotechnology.

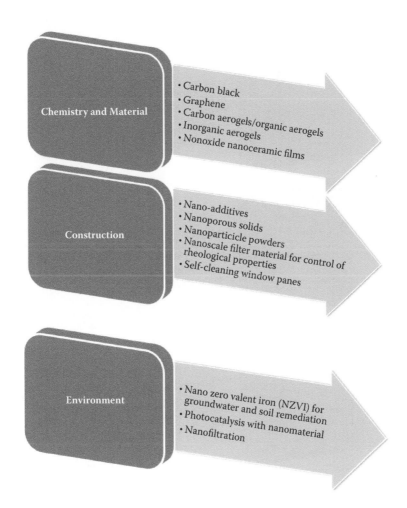

FIGURE 1.6 (*Continued*)
Emerging opportunities in nanotechnology.

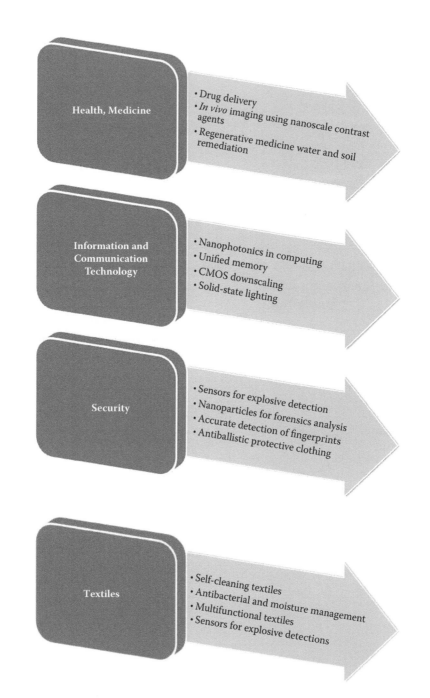

FIGURE 1.6 (*Continued*)
Emerging opportunities in nanotechnology.

FIGURE 1.7
Commercial nanogalaxy.

been forecasted to reach 1.3 billion pounds (£) by 2015. Global demand for nanocomposites in electronics and electrical end-use is expected to surge at a CAGR of more than 14% through 2015.[7]

The last five decades of nanotechnology since its conception have been a period of rapid developments in the science of nano with knowledge augmentation and a simultaneous thrust in the search for demonstrable applications. The next nano-steps will challenge every neuron of the human brain to convert today's fiction into tomorrow's reality in reproducible, inexpensive, and cost-effective ways to provide affordable and useful products to our society. Although the nanoworld of today with its structures, devices, and systems including microscopic procreating robots is being driven by the slogan "smaller, faster, better," there are untold stories of the nanoworld involving nanosafety yet to unfold that could take the world by surprise that may act as barriers to nano-commercialization.

Driving the nano to success will require a constructive cross-disciplinary confluence of all facets of the human mind, evolving skills and expertise to contemplate, foresee, and address as many social, legal (including intellectual property rights), cultural, ethical, religious, philosophical, and political implications of the nanoworld of tomorrow and the days after.

The evolving nano-economics (E) will be strongly influenced by the quality of concepts (C_1), application of novel and cost-effective manufacturing methods (M), and the rate of commercialization (C_2), wherein $E \sim fn\ M\ .C_1^n.C_2^m$ where in the variable n is dependent on the impact factor of the concept and m is dependent on the mode of commercialization.

The rate of commercialization (C_2) is strongly influenced by several inter-dependent market-linked factors, such as the following:

1. Characteristics of the nanovation, its ease of manufacture, and the range of applications it can be put to directly as a product, be incorporated into other products, or enable creation of useful products/services

2. Interoperability

3. Positioning vis-à-vis competing products/services

4. Ease of getting regulatory clearances

5. Product life cycle

6. Stage and manner in which it is IPR-protected in different jurisdictions

7. Nature of disclosures made—how, when, and to whom

8. Techno-legal proceedings in IPR offices (opposition or revocation proceedings on IPR infringement, validity challenge, enforcement against infringers including settlement of the techno-legal proceedings through strategic arrangements)

9. Value of IP

10. IP transactions done with the IPR portfolio (acquiring, selling, licensing, cross-licensing, assigning, sharing of trade secrets and know-how, etc.)

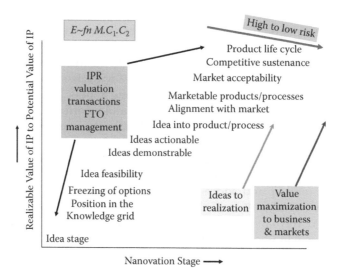

FIGURE 1.8
Nanovation—IPR.

11. Institutional mergers, acquisitions, setting up of joint ventures, cooperative marketing arrangements, etc. including nature and dynamics of collaboration

12. Business model being employed for commercialization

The realizable value of IP to the potential value of IP also varies and is influenced by a number of factors some of which are illustrated in Figure 1.8. As the nanovation approaches market expectations, the realizable value to the potential value increases as the business takes on lower risks. A very important aspect is a reasonable analysis of an invention vis-à-vis a set of relevant patents to assess the freedom to operate in specific jurisdictions of business interest. However, much of the value is strongly dependent on the manner in which the IPR is managed from concept stage to the contemplated stage of commercialization based on the business model being adopted.

1.1 Evolving Patents Landscape in Nanotechnology

The ObservatoryNANO constituted by the European Commission publishes an annual report on statistical patent analysis dealing with nanotechnology. These reports can be accessed on their website at http://www.observatorynano.eu/. The latest report presents the trends in global patent filings in nanotechnology in general and then proceeds to analyze them in terms of the technical areas of their application and the overall contribution to these global trends from various countries. Figure 1.9 shows the global trends. It is clear that there was steady growth in patent filings in nanotechnology from 1990 to 2000 as the world was coming to grips with

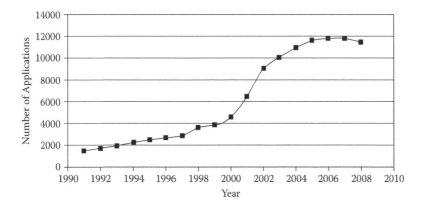

FIGURE 1.9
Patent applications in nanotechnology (Observatory NANO Factsheets, March 2011).

the nuances of nanotechnology with emerging indications of promising applications. The period of 2000 to around 2006 saw rapid growth in patent applications as investments in R&D for "nano" were enhanced and industries started sprouting in various parts of the world to exploit the promise of nanotechnology. From 2004 onward, a sense of realism began to set in and various business models in the commercialization of nanotechnology started to emerge. Patent litigations that were initiated in early 2000 were being decided in courts followed by various types of settlements between the litigating parties, and patent offices in various parts of the world were attaining rudimentary expertise to examine patent applications related to nanotechnology. Such factors coupled with rising costs, global economic recession, and increasing risks have brought in a sense of caution leading to a gradual plateau in patent filings since 2005.

Figure 1.10 shows the contributions of various countries to global patenting in nanotechnology. The field continues to be dominated largely by the United States (47%), Japan (29%), and the European Union (20%) since the early stages of the development of nanotechnology with insignificant contributions from other countries. Within the EU, Germany (8% of global filings), France (4%), and the United Kingdom (4%) have taken lead positions. However, in recent years, China and Korea (already 4% of global filings) have begun to make their presence in the global nanotechnology-patenting scene known and promise to pose as major competitors in the future global marketplace. The global patenting trends and dominance are very similar to the number of publications in the field of nanotechnology from these countries.

As seen in Figure 1.11, the European study shows that patenting in the semi-conductor field has been the highest followed by medical science, hygiene,

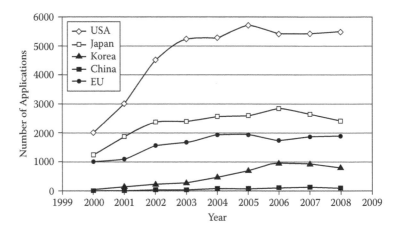

FIGURE 1.10
Major contributors to the global nanotechnology patenting landscape convergence (Observatory NANO Factsheets, March 2011).

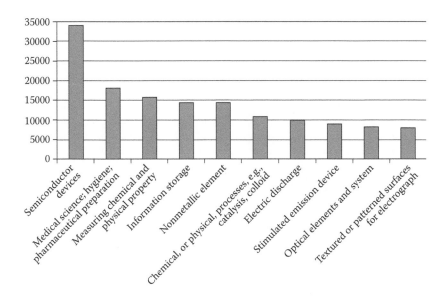

FIGURE 1.11
Patent filing trend (Observatory NANO Factsheets, March 2011).

and pharmaceutical preparations. Progress in the science and technology of nanotechnology would need monitoring and measuring of physical-chemical properties of such small materials, entities, etc. and, hence, has the next level of scores in the filing scenario. Energy storage has been a thrust area of research and nanoscience is expected to play a major role; therefore, patents in this field in such high numbers are not surprising. The other fields in which nanotechnology-related patents have been filed in large numbers mirror the developments in those fields, clearly signaling the relevance of nanotechnology as thrust area in such applications.

The OECD report has a list of companies (Figure 1.12) in Europe that are the largest patent applicants in nanotechnology. Among the list of electronics, chemical, and high specialty materials companies, a consumer and cosmetics company such as L'Oreal indicates the role of nanomaterials in such consumer-driven applications.

1.2 Trends in the United States

In a presentation made by Bruce Kisliuk[8] of the U.S. Patent and Trademark Office (USPTO) in September 2010, he indicated that USPTO was applying a two-prong test to classify patents with "nano" as subject matter, namely, at

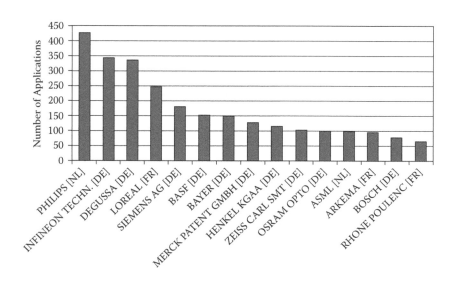

FIGURE 1.12
Patent filing trend. (Observatory NANO Factsheets, March 2011.)

least one physical dimension of ~1 to 100 nm and special property, function, or effect uniquely attributable to the nanosized dimension.

The subject matter for the filings of patents in nanotechnology as presented by Kisliuk was nanostructure (33%); manufacture, treatment, or detection of nanostructures as a group (29%); specified use of nanostructures (27%); mathematical algorithms, for example, specially adapted for modeling configurations or properties of nanostructures (<1%); and miscellaneous (<1%).

The data obtained by the authors from the USPTO in September 2011 indicate a steady rise in the number of patent filings in nanotechnology (Figure 1.13). Although in absolute terms the filings by U.S. inventors in USPTO have remained consistently higher than those filed by foreign inventors, the percentage of filings by U.S. inventors of the total filings has began to show a marginal downward trend since 2007. In contrast, the percentage of filings by foreign inventors of the total filings continues to show a steady rise since 2007. These trends clearly indicate that the world is fast catching up in terms of inventions in nanotechnology and the United States soon could be losing its dominance in the nanotechnology patent landscape.

Kisliuk's presentation indicated that Japan tops the list of foreign filers with a contribution of ~45%, followed by Germany (~10%), South Korea (~8%), France and Canada (~5% each), Taiwan (~4.5%), United Kingdom and Switzerland (~3.5% each), and Israel and Italy (~2% each).

An analysis of the 6245 patents granted in U.S. Class 977 (nanotechnology) until around July 2010 in various technology centers of the USPTO shows the following:

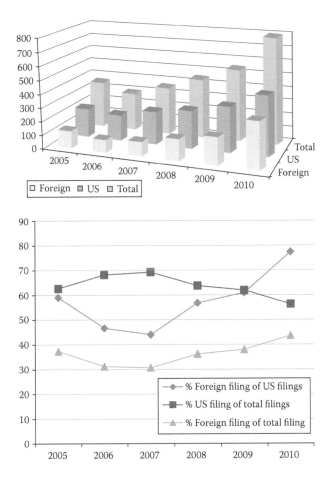

FIGURE 1.13
Top: Nanotechnology patent filings by U.S. and foreign applicants in the United States. Bottom: Comparison of nanotechnology patents by U.S. and foreign applicants.

	Technology Center	No. of Patents
TC1600	Biotechnology and organic chemistry	972
TC1700	Chemical and materials engineering	1813
TC2100	Computer architecture software and info security	22
TC2600	Communications	242
TC2800	Semiconductors, electrical, optical systems	2737
TC3600	Transportation, construction, electronic commerce	51
TC3700	Mechanical engineering, manufacturing and products	408

Data obtained by the authors in September 2011 from the USPTO on the first named assignees in the United States with at least 10 granted patents are presented in Table 1.1. It is interesting to find a few universities such as

TABLE 1.1

U.S. Patents Granted in Class 977 between 2005 and June 2011

First-Named Assignee	2005	2006	2007	2008	2009	2010	2011 (Through June)	All Years
Samsung Electronics Co., Ltd.	9	9	10	12	22	42	27	131
International Business Machines Corporation	7	7	11	12	28	28	21	114
Hewlett-Packard Development Company, L.P.	14	8	18	13	13	15	7	88
Nantero, Inc.	5	7	14	8	17	27	9	87
Intel Corporation	10	9	11	12	12	9	5	68
William Marsh Rice University	11	11	7	11	10	10	6	66
University of California	12	6	10	3	12	7	3	53
Nanosys, Inc.	1	6	5	10	9	15	2	48
Tsinghua University	2	1	6	1	9	19	7	45
Canon Kabushiki Kaisha	7	7	4	5	5	7	3	38
Industrial Technology Research Institute, Taiwan	6	5	1	3	10	9	0	34
Massachusetts Institute of Technology	2	3	6	7	5	6	4	33
Micron Technology, Inc.	2	6	2	3	5	11	3	32
Toshiba Corporation	2	1	2	6	5	8	5	29
3M Innovative Properties Company	0	3	4	7	7	5	2	28
Northwestern University	1	0	2	7	6	10	1	27
Samsung Sdi Co., Ltd.	3	2	0	5	7	8	1	26
California Institute of Technology	4	0	5	4	1	6	5	25
Fujitsu Limited	0	2	2	5	3	6	6	24
General Electric Company	3	4	4	4	4	4	1	24
Japan Science and Technology Agency	2	3	3	3	3	6	4	24

TABLE 1.1 (*Continued*)

U.S. Patents Granted in Class 977 between 2005 and June 2011

First-Named Assignee	2005	2006	2007	2008	2009	2010	2011 (Through June)	All Years
United States of America, National Aeronautics and Space Administration	2	2	5	0	6	5	4	24
United States of America, Navy	4	1	0	5	5	8	1	24
Infineon Technologies Ag	5	1	3	1	5	8	0	23
Sony Corporation	2	2	2	3	4	8	2	23
E. I. Du Pont De Nemours and Company	1	1	4	5	4	4	3	22
Harvard College	0	1	6	3	4	2	3	19
D-Wave Systems Inc.	11	4	1	2	0	0	0	18
Hitachi, Ltd.	6	3	3	0	2	4	0	18
Stanford University	3	2	3	5	1	2	2	18
Ut-Battelle, LLC	2	3	4	2	0	4	2	17
Hon Hai Precision Ind. Co., Ltd.	3	0	2	3	6	0	1	15
Hyperion Catalysis International, Inc.	3	3	0	2	1	0	6	15
Research Foundation of State University of New York	2	3	2	1	4	3	0	15
Commissariat A L'energie Atomique	1	0	3	3	2	3	2	14
Freescale Semiconductor, Inc.	0	1	5	4	0	3	1	14
Fuji Xerox Co., Ltd.	4	1	1	2	3	3	0	14
Samsung Electro-Mechanics Co., Ltd.	0	0	1	0	3	6	4	14
Seagate Technology, LLC	2	0	1	3	1	1	5	13
Sharp Laboratories of America, Inc.	0	1	3	5	2	1	1	13
University of Oklahoma	3	3	1	3	2	1	0	13
University of Central Florida	0	1	0	3	0	6	2	12
Applied Nanotech Holdings, Inc.	0	0	0	2	3	4	2	11

(*Continued*)

TABLE 1.1 (*Continued*)

U.S. Patents Granted in Class 977 between 2005 and June 2011

First-Named Assignee	2005	2006	2007	2008	2009	2010	2011 (Through June)	All Years
NEC Corporation	3	1	2	0	4	0	1	11
Penn State Research Foundation, Inc.	2	3	1	3	0	2	0	11
Qunano AB	0	0	0	3	1	4	3	11
STMicroelectronics S.R.L.	0	0	1	1	2	4	3	11
Battelle Memorial Institute	2	3	1	0	0	3	1	10
Eastman Kodak Company	0	0	1	2	5	0	2	10
Honda Giken Kogyo Kabushiki Kaisha (Honda Motor Co., Ltd.)	0	1	1	2	3	1	2	10
National Institute of Advanced Industrial Science and Technology	2	0	0	0	1	4	3	10
Panasonic Corporation	0	0	0	1	4	5	0	10
Rensselear Polytechnic Institute	1	0	2	0	4	2	1	10
Showa Denko Kabushiki Kaisha	0	1	0	2	2	3	2	10
SII Nanotechnology, Inc.	4	0	1	2	2	0	1	10
University of Texas	2	0	1	1	2	2	2	10

Rice University, University of California, and Tsinghua University, Industrial Technology Research Institute, Taiwan, figuring in the top slots. The presence of some universities in the top patentees list also reaffirms that selected universities have focused on intense nanotechnology R&D and are expected to be key participants in IP transactions in due course of commercialization of their technologies. Further, the list also contains several Japanese and Korean companies that have shown aggressive patenting in the field of nanotechnology; they are likely to take lead positions in the United States.

Several universities and R&D centers figure in the list of assignees of U.S. patents in Table 1.1 that over the years have developed into centers of excellence in nanotechnology. With their rich patent portfolios and specialized knowledge, these centers are expected to lead activities related to upstream knowledge sharing to enrich domain knowledge, strategic partnering/collaborative working, technology transfer, and IP transactions. In early 2000, one has seen the sprouting of several technology start-ups by innovators, some of which survived over time, some have been taken over by large corporations, with many others dwindling off for a combination of diverse reasons.

The "nanoscape" hints beyond doubt that IPR management has to become an integral part of "nanoventing" to enable facile commercialization of "nanovations."

References

1. Taniguchi, N., On the basic concept of nano-technology, *Proc. Intl. Conf. Prod. Eng. Tokyo*, Part II, Japan Society of Precision Engineering, 1974.
2. Liu, L., ed., *Emerging Nanotechnology Power—Nanotechnology R&D and Business Trends in the Asia Pacific Rim*, World Scientific Publishing Co. Pte. Ltd., 2009.
3. WIPO (World Intellectual Property Organization), 2011. World Intellectual Property Report 2011, The Changing Face of Innovation, http://www.wipo.int/export/sites/www/econ_stat/en/economics/wipr/pdf/wipr_2011.pdf
4. Palmberg, C., Dernis, H., and Miguet, C., Nanotechnology: An overview based on indicators and statistics, *OECD Science, Technology and Industry Working Papers*, No. 2009/07, OECD Publishing. 2009. doi: 10.1787/223147043844.
5. Finardi, U., Time-space analysis of scientific citations in patents: Science-innovation links in the nano-technological paradigm, paper presented at the 6th Annual Conference of the EPIP Association Fine-Tuning IPR Debates, September 2011, http://www.epip.eu/conferences/epip06/papers/Parallel%20Session%20Papers/FINARDI%20Ugo.pdf
6. The Project on Emerging Nanotechnologies, http://www.nanotechproject.org/inventories/consumer/analysis_draft/
7. Electronics.ca publications, Electronics industry market research and knowledge network, http://www.electronics.ca/presscenter/categories/Nanotechnology/
8. Kisliuk, B., Nanotechnology Partnership Forum NIST, Gaithersburg, MD, September 13, 2010, http://www.nist.gov/tpo/upload/BruceKisliuk.pdf

2

Patents: A Background

The tools of intellectual property rights (IPR) of relevance to nanotechnology are patent, trademark, copyright, and industrial design registration. A patent protects inventions and hence this tool is of particular significance to nanotechnology. Inventions cover disruptive technical developments and incremental technical improvements in processes and products including any substance or material, device, machine, or apparatus.

In operational terms, a patent is a grant by a sovereign or state to a person giving him or her exclusive right to stop others from making, using, exercising, and vending his or her invention for a limited period of time, in exchange for disclosing his invention in a patent specification in a manner such that a person skilled in the art can reproduce the invention without undue experimentation. Thus, a patent is effectively a negative right giving the patent holder the right to exclude others from using his or her patented invention for a limited period of time. However, it must be appreciated that the patent holder has a qualified right to use his or her invention. The patent system requires a disclosure of the invention in a specified manner (patent specification) defined by the statute.

The patent system gives the patent holder a monopoly on his or her invention for a limited period, but clearly discourages any monopolistic behavior that may jeopardize competition in the marketplace. The control on monopolistic behavior or overuse of patent rights are built into the patent laws and brought into effect through competition/antitrust laws in various jurisdictions.

Patent applications and granted patents are well-acknowledged indicators of national innovative capacity in developed market economies.[1] Public and private entities across the globe are convinced of nanotechnology's potential and are staking their claims. International rivalries are growing, political alliances are forming, and battle lines are being drawn.[2]

In the era of "knowlitics®" (knowledge politics), patents provide the invisible protection boundaries that have already started to get diffused with closely overlapping claims in the crowded terrain of nanotechnology. Such a dense patented landscape has led to drawing of strategic battle lines, alliances, mergers, acquisitions, joint ventures, etc. to gain and retain the needed competitive edge.

It is of paramount importance that patents are drafted with a business perspective so that the patent application and subsequently the granted patents are strong enough to be defended when challenged and sufficiently clear in terms of the boundary and area of protection so that the claimed invention can be enforced against infringers. Further, a strong patent with strategically crafted claims can facilitate settling of litigation between litigating partners, striking collaborations, sharing of IPR, etc.

Patent rights are territorial, which means that if a patent is granted in the United States, the patent holder's right is limited only to the United States. If a patent for the same invention was not granted in Canada, then the U.S. patent holder has no patent rights in Canada and, therefore, anyone would be free to use the invention in Canada without requiring prior authorization from the U.S. patent holder. Thus, one has to apply for the patent in various countries and have them granted in those countries to attain any patent rights on the invention in those countries. Further patents have a limited term of 20 years from the date of filing the patent application in a country and, after grant, the patent holder has to pay annuity fees to retain the rights. Nonpayment of the annuity fees within a specified period results in forfeiting the patent right in that jurisdiction.

The Trade Related Aspects of Intellectual Property Rights Agreement (TRIPs) within the World Trade Organization (WTO) framework has defined the minimum standards for patentability (Article 27 of the TRIPS Agreement). However, there are flexibilities in the definition of "patentable subject matter" as the Member States may exclude certain types of inventions from being patentable under their jurisdiction so long as they satisfy the minimum conditions laid down by the TRIPS Agreement.

Article 27 of the TRIPS Agreement titled "Patentable Subject Matter" states:

- Subject to the provisions of paragraphs 2 and 3, patents shall be available for any inventions, whether products or processes, in all fields of technology, provided that they are new, involve an inventive step and are capable of industrial application. Subject to paragraph 4 of Article 65, paragraph 8 of Article 70 and paragraph 3 of this Article, patents shall be available and patent rights enjoyable without discrimination as to the place of invention, the field of technology and whether products are imported or locally produced.

- Members may exclude from patentability inventions, the prevention within their territory of the commercial exploitation of which is necessary to protect *ordre public* or morality, including to protect human, animal or plant life or health or to avoid serious prejudice to the environment, provided that such exclusion is not made merely because the exploitation is prohibited by their law.

- Members may also exclude from patentability:
  - diagnostic, therapeutic and surgical methods for the treatment of humans or animals;
  - plants and animals other than micro-organisms, and essentially biological processes for the production of plants or animals other

than non-biological and microbiological processes. However, Members shall provide for the protection of plant varieties either by patents or by an effective *sui generis* system or by any combination thereof. The provisions of this subparagraph shall be reviewed four years after the date of entry into force of the WTO Agreement.

For the purposes of this Article, the terms "inventive step" and "capable of industrial application" may be deemed by a Member to be synonymous with the terms "nonobvious" and "useful," respectively.

The benchmarks for patentability are illustrated in Figure 2.1.

- An invention must be patentable subject matter, simultaneously satisfying benchmarks of
  - Novelty (be novel)
  - Inventive step (nonobvious to a person of ordinary skill in the art)
- Utility/capable of industrial application (useful) and sufficiently described to enable one skilled in the art to make and use the invention

In view of this, patentable subject matter could vary from jurisdiction to jurisdiction and could have impact on patentability of inventions in nanotechnology.

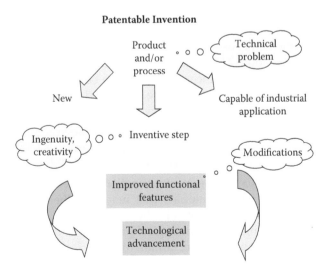

FIGURE 2.1
Benchmarks for patentability.

2.1 Patentable Subject Matter

The issues of patentability related to subject matter of nanotechnology could span from exclusions in different jurisdictions based on the national patent law and the interpretation of the following phrases:

- exclude from patentability inventions, the prevention within their territory of the commercial exploitation of which is necessary to protect *ordre public* or morality, including to protect human, animal or plant life or health or to avoid serious prejudice to the environment
- diagnostic, therapeutic and surgical methods for the treatment of humans or animals
- plants and animals other than microorganisms, and essentially biological processes for the production of plants or animals other than nonbiological and microbiological processes

Serious debates involving nanotechnology have already taken center stage. The issues that may bring in these features to question the patentability of inventions in nanotechnology could include the following:

- the classification of findings in nanotechnology as discoveries or inventions especially that involve products resulting from manipulation of naturally occurring substances to nano dimensions even if they exhibit surprising properties as that would possibly qualify as discovering new properties of existing and known materials or products of nature
- items related to nano materials being employed in nanobiotechnology, nanomedicine, use of self-scaffolding materials for the creation of a new generation of biomaterials and tissues, methods of introducing nontherapeutic enhancement of human beings, use of nanomaterials in stem cell research to create structured-functionalized nano surfaces to mimic the human embryo, etc. (*ordre public* or morality)
- use of nano-biological processes (not microbiological processes) to modify genetic identity of human beings and animals (ethical issue and bypassing one of the exclusions of Article 27 (3)(b))
- use of nanotags exploiting miniaturization of intelligence and ICT for incorporation into materials that may be either implanted in the human body or introduced in the human body to derive in situ information of ongoing the biological processes in the human body (issue of privacy, surveillance, human dignity)

Debates are also beginning to surface in trying to differentiate therapeutic methods from cosmetic methods, methods of treatment and surgery involving new-generation nanotechniques, etc., causing a blurring of boundaries between methods of treatment and diagnostics practiced on the human body. Such issues will be on the rise and inventions in nanotechnology will be debated on a broader canvas to create an effective ethics-based regulatory framework to facilitate progress with appropriate protection of inventions in these areas so that investments continue to fund creative advancement and businesses.[3]

2.2 Novelty (New)

An invention is considered new if it does not form part of state of art. The meaning of "state of art" or "prior art" in the context of an invention is that the "invention in whole" is publicly accessible information anywhere in the world prior to the filing of the first patent application for the particular invention.

It is to be noted that for consideration of novelty, one is not allowed to combine separate documents or information on parts of the invention available in different sources to reconstruct the invention. To establish lack of novelty, the subject matter of the invention needs to be disclosed in a single document or in a single source without needing to combine various documents and different sources of information.[4]

The information provided in the prior art document needs to be sufficient for a person of ordinary skill in the art to practice the invention based on the disclosure in the document, combined with the available general knowledge at the point in time when the document was published. This feature is illustrated in Figure 2.2.

What Is New?

Invention does not form part of existing

Relevant

Knowledge

Any publication information in any form accessible to public

Journal, conference proceedings, brochure, catalog, etc.

Available to Public

Anywhere in the WORLD

Prior to First Filing of Patent Application

FIGURE 2.2
Novelty aspects.

It is important to note that the invention may suffer from loss of novelty even if the prior art mentions related matter from which the invention or aspects of the invention can be inherently derived directly or unambiguously.

With reference to nanotechnology, one has to identify at least one clear difference in the physical properties between two products. For example, if a product has a fluid-conducting channel that is 1 μm in width and another product has a fluid-conducting channel that is 1 nm in width, there cannot be anticipation. The product with the channel that is 1 nm in width is novel with respect to the product that has a channel that is 1 μm in width. However, under the doctrine of inherency, a prior art reference may "inherently" anticipate a claimed invention, even if the reference does not expressly disclose the later invention. Referring to the channel example, suppose the prior art reference did not clearly indicate the exact physical size of the fluid-conducting channel. What could be the arguments on anticipation? In this case, one may argue that the disclosed fabrication method in the prior art to form the 1 μm conducting channel "inherently" would have produced a channel 1 nm in width and, therefore, the channel of 1 nm width is anticipated.

Some nanotechnology inventions, however, involve nanoscale formulations of previously disclosed chemical compounds, structures, and materials. In the case where nanoscale inventions exhibit properties that are, in some measure, unanticipated or different from those found in larger scale prior art, exceptions have been made.

For example, in *BASF v. Orica Australia*,[5] the European Patent Office's (EPO's) Technical Board of Appeals (TBA) held that

- A prior patent which disclosed polymer nanoparticles larger than 111 nm did not destroy the novelty of a subsequent application by Orica for nanoparticles smaller than 100 nms.

- Orica's smaller particles exhibited remarkably improved technical properties resulting in a glossier coat compared to the larger particles protected under the prior patent. The difference in properties was held to be sufficient to impart novelty.

Generally, even the slightest overlap is sufficient to destroy novelty, but exceptions have been applied liberally to nanoscale inventions.[6] Under the EPO's approach to assessing novelty, the overlap must be a narrow relative to the larger prior art range, sufficiently far removed from the larger range, and indicative of an invention.

For example, in a case between SmithKline Beecham Biologicals and Wyeth Holdings Corporation,[7] the question was whether SmithKline's patent application on a hepatitis B vaccine adjuvant* lipid measuring 60 to 120 nm lacked novelty in light of a prior patent on a similar adjuvant with particles measuring 80 to 500 nm.

---

* An adjuvant is a pharmacological or immunological agent often included in vaccines to enhance the recipient's immune response to a supplied antigen.

The TBA found that SmithKline's patent was novel because the overlap was (a) narrow—only 10% of the larger range in the earlier patent; (b) at the extreme lower end of the prior art range; and (c) exhibited significantly improved adjuvancy—the smaller particles resulted in an unexpected and favorable shift in immune response.

The prior art gave little guidance on how to prepare the smaller particles. A skilled person who followed the vaccine supplier's protocol would have produced particles of between 115 and 951 nm. The technical teachings in the prior art, therefore, were not considered relevant to SmithKline's patent application.[6]

The nanotechnology patent landscape is already a dense web of overlapping rights and, hence, a detailed prior art search and analysis is necessary before embarking on any project. The challenge is to overcome the claims in the prior art[6] to optimize the available resources and execute R&D in a systematic manner.

It is often observed that inventions are published unknowingly and unintentionally out of ignorance or low awareness of the fundamentals of the patents. Proper care must be observed before publishing the subject matter related to an invention as it may lead to invalidating patentability of one's own invention because of such a publication. Some of the ways in which novelty is destroyed are as follows:

- Advertisements, marketing brochures
- Public handouts, preproduct launch publicity including press releases, annual reports, and announcements on the Web
- Drawings and product details for development to vendor/contract manufacturers without a nondisclosure agreement
- Disclosure at the time of testing product to external agencies (e.g., for certification) that may be accessed by the public
- Sharing information with would be or probable collaborators (without a nondisclosure agreement)
- Technology details/data provided to market research agencies without nondisclosure instructions before product launch
- Interviews on television, in newspapers and magazines, etc.
- Public testing of inventions without adequate cover
- Disclosures in questionnaires for collection of information from the marketplace without specific confidentiality commitments
- Publications in journals and books, lecture notes for circulation without specific instructions on confidentiality
- Nonconfidential laboratory manual/brochure
- Funding proposal submitted to various agencies that does not have any restrictive clauses on confidentiality
- Seminars, conferences, and publication in proceedings

2.3 Inventive Step (Nonobviousness)

Novelty exists if there is a difference between the invention under consideration and relevant prior art. The invention has an inventive step if said invention is not obvious to a "person of ordinary skill in the art" in light of the prior art.

It is to be noted that with regard to the phrase "person skilled in art," the word "art" is used in a broad context of the field of invention or the technological area and required skill sets related to the invention and is not limited to a specific branch or stream of engineering or science. "Art" should not be construed to mean aesthetics.

Various judgments have interpreted the term "person skilled in the art" as follows:

- The person skilled in the art need not be (is not suppose to be) an expert in the art. He or she could be an ordinary practitioner aware of what was common general knowledge in the art at the relevant date.
- The person is supposed to be aware of common general knowledge at the corresponding time (date) of the prior art/state of art. The relevant documents should have been accessible to this person. He or she would have had access to everything in the "state of art," in particular the documents cited in the search report.
- The person would have had at his or her disposal the normal means and capacity for routine work and experimentation.
- The person would be of average skill not engaged in creative thinking.
- Based on the technological field, it may be assumed that person skilled in the art is a group of people and not necessary a singular person.

Interestingly the person skilled in the art is expected to react in a way common to all skilled persons at any time, that is, based on facts, identifying possible obstacles that could impede the successful realization of the solution to a defined problem. The knowledge of the notional person skilled in the art has to be considered as that of a team of appropriate specialists that appreciate the difficulties still to be expected when considering a solution in the context of the problem. However, the skilled person has to be assumed to lack the inventive imagination to solve problems for which routine methods of solution did not already exist.

Broadly speaking, the perception of a notional person skilled in the art as to the obviousness of the invention is used as a determinant. Considerations in various jurisdictions are as follows:

- The person skilled in the art, having regard to any item of prior art or common general knowledge, would have arrived at the claimed invention (Europe).

- It is not obvious to a person skilled in the art, with regard to the matter that was in the state of art, that is, was available to the public, before the priority date of the invention (United Kingdom).
- Any item of prior art or common general knowledge would have motivated a person skilled in the art to reach the claimed invention (Japan).
- Any item of prior art or common general knowledge would have motivated, with a reasonable expectation of success, a person skilled in the art to reach the claimed invention (United States).[8]

The invention is obvious with respect to prior art if there is a lack of ingenuity shown by the inventor, if the invention does not contribute to substantial technical advancement or economic significance (or both), and it is only a logical extension of what is provided in the prior art. It is important to consider if the invention would occur to a person skilled in the art. That means it is necessary that the person skilled in the art needs to make substantial effort and use ingenuity beyond a normal application of skill to arrive at the invention.

In case of nanotechnology inventions, generally, an invention is considered obvious if it miniaturizes known elements, performing the same function, and yields no more than might be expected from the diminished size. It is to be noted that making something smaller does not automatically make it new or inventive. Miniaturization-based inventions should always demonstrate an enhanced technical effect derived from the size. Technology is considered nonobvious if it produces new and unexpected results or serves previously unrecognized functions that overcome a technical problem related to the prior art. As practically all nanoscale technologies display these characteristics, only those results that are not likely to emerge from extrapolations by a skilled person working with smaller structures are deemed patentable.[6]

In the *SmithKline Beecham Biologicals v. Wyeth Holdings Corporation* case, the vaccine adjuvant was held to be inventive because of its unexpectedly improved effect and the fact that nothing in the prior art had suggested that a skilled person might consider reducing the particle sizes to achieve the desired advantage.

In *BASF v. Orica Australia*,[5] Orica's claimed invention involved manufacturing polymer particles at 100 nm or less by initiating polymerization at temperatures below 40°C.

BASF argued that the invention was obvious because a prior patent had disclosed the same manufacturing process using temperatures below 50°C to yield particles averaging 111 nm or more.

BASF argued that a skilled person exercising no inventive effort and repeating reactions on a trial-and-error basis for all temperatures between 0°C and 50°C would have derived sub-100 nm particles at temperatures below 40°C.

The EPO rejected this argument and reasoned that the prior patent suggested using temperatures not exceeding 50°C. While this "did not rule out the use of temperatures below 40°C, it was far from suggesting their use."

The patent was aimed at manufacturing particles larger than 111 nm only. A skilled person following the teachings of the prior patent would not have used temperatures below 40°C or foreseen that lower temperatures would result in particles smaller than 100 nm.

The TBA held that Orica's invention provided, for the first time, a method of creating smaller variants of polymer nanoparticles and was, therefore, inventive.[6]

2.3.1 Invention to Be Considered as a Whole

While establishing the aspect of obviousness with respect to prior art, the invention under consideration (for which patentability is to be established) is to be considered in totality. The invention could be the cognitive effect of certain components or elements and related features of the article/device. Those components or elements are not allowed to be deciphered and evaluated individually for obviousness/inventive step consideration.

A set of tests for inventive step (nonobviousness) is illustrated in Figure 2.3. The invention is to be identified. The relevant prior art needs to be searched. The closest prior art has to be studied. The next set of related prior art has to be analyzed. One has to assess whether the teachings of the prior art singly or in combination would lead a person skilled in the art to the invention without undue experimentation.

The matters related to inventive step (nonobviousness) have been the subject of varied interpretations by patent attorneys/agents, patent granting authorities, courts, appellate boards, etc. Inventive step in various jurisdictions has been worded differently and further interpreted by courts in a variety of ways.[8]

2.3.2 Europe

In the European Patent Office (EPO), as per Article 56 European Patent Convention (EPC), an invention is considered to involve an inventive step if, having regard to the state of the art, it is not obvious to a person skilled in the art. Article 54(2) and (3) EPC defines the state of the art for the purposes of considering an inventive step. Thus, according to EPO the question to consider, in relation to any claim defining the invention, is whether at the priority date of that claim, having regard to the art known at the time, it would have been obvious to the person skilled in the art to arrive at something falling within the terms of the claim. If so, the claim is bad for lack of an inventive step. The term *obvious* means that which does not go beyond the normal progress of technology, but merely follows plainly or logically from the prior art; that is, something that does not involve the exercise of any skill or ability beyond that to be expected of the person skilled in the art.

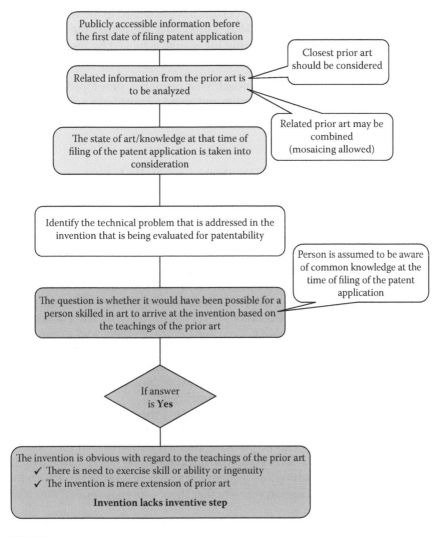

FIGURE 2.3
Tests for inventive step.

The Boards of Appeal of EPO normally apply the "problem and solution approach." Essentially this consists of the following:

1. Identifying the closest prior art
2. Assessing the technical results (or effects) achieved by the claimed invention when compared with the closest state of the art established
3. Defining the technical problem to be solved as the object of the invention to achieve these results

4. Examining whether a skilled person, having regard to the state of the art in the sense of Article 54(2), would have suggested the claimed technical features for obtaining the results achieved by the claimed invention

The problem and solution approach was primarily developed to ensure objective assessment of inventive step and avoid ex post facto analysis of the prior art.

2.3.3 United States

Nonobviousness according to U.S. Patent Law 35 U.S.C §103 requires that everything that falls within the scope of the claim must be inventive. Almost 150 years ago, the Supreme Court in *Hotchkiss v. Greenwood* observed that a patent should only be awarded for those inventions that embody a "degree of skill and ingenuity" beyond that of "an ordinary mechanic acquainted with the business." During the last five decades to arrive at decisions related to non-obviousness, the U.S. courts have considered a combination of questions such as (1) the scope and content of the prior art; (2) the differences between the prior art and the patent claim as a whole; (3) the level of ordinary skill in the pertinent art; and (4) the presence of secondary considerations, such as commercial success, long-felt but unresolved needs, the failure of others, and expert testimony. U.S. courts have also considered the motivation–teaching–suggestion (MTS) test, which requires that there must be a specific finding of a motivation, teaching, or suggestion in the prior art to combine references in the manner claimed to support an obviousness conclusion.

In the KSR matter (*KSR Int'l Co. v. Teleflex, Inc.*), the patent at issue was U.S. Patent No. 6,237,565B1, entitled "Adjustable Pedal Assembly with Electronic Throttle Control," wherein the patentee claimed a mechanism for combining an electronic sensor with an adjustable automobile pedal so the pedal's position could be transmitted to a computer that controls the throttle in the vehicle's engine. The question was whether there was an inventive step with regard to the prior art.

The Supreme Court in *KSR*, while reversing the judgment of the Federal Circuit, also questioned the lower court's narrow conception of the obviousness inquiry consequent in its application of the MTS test. While arriving at the decision that there was no inventive step, the Supreme Court observed that while considering the question of inventive step, broad searching would be the normal expectation of the skilled worker; secondary evidence was crucial in assessing inventiveness; the pool of prior art that is relevant is much broader than that of the particular problem at hand; any prior art information belonging to a "field of endeavor" was relevant, thereby seemingly permitting, for example, the combination of information not clearly related to the invention with the knowledge of the skilled worker to render an invention as obvious; and the skilled worker is expected to combine information from multiple documents like pieces of a puzzle to obtain a solution to a problem.

2.3.4 United Kingdom

In the United Kingdom, on matters related to inventive step, Section 1(1)(b) of the U.K. Patents Act 1977 states that a patent may be granted only if it involves an inventive step. Section 3 states that an invention should be considered to involve an inventive step if it is not obvious to a person skilled in the art, having regard to matter which was in the state of art, that is, was available to the public before the priority date of the invention. The courts in the United Kingdom traditionally have formulated a four-step approach to assessing obviousness:

1. Identify the claimed inventive concept.
2. Identify the common general knowledge known to a skilled but unimaginative addressee in the art at the priority date.
3. Identify the differences, if any, between the matters identified as known or used and the alleged invention.
4. Decide whether viewed without any knowledge of the alleged invention those differences constitute steps that would have been obvious to the skilled individual or whether they require any degree of invention.

All claimed members of the known class must have the advantage and the advantage must not be one that those skilled in the art would expect to find in a large number of the previously disclosed class.

The key legal precedent in the United Kingdom is the four-step Windsurfing test set out by the Court of Appeal in *Windsurfing International Inc. v. Tabur Marine (Great Britain) Ltd,* [1985] RPC 59. Further questions to be taken into account were laid out in *Haberman v. Jackal* [1999] FSR 685 (at 699 to 701).

Lord Hoffmann, in *Biogen Inc. v. Medeva plc* [1997] RPC 1 (at page 34) made the following observations:

> Whenever anything inventive is done for the first time it is the result of the addition of a new idea to the existing stock of knowledge. Sometimes, it is the idea of using established techniques to do something which no one had previously thought of doing. In that case the inventive idea will be doing the new thing. Sometimes it is finding a way of doing something, which people wanted to do but could not think how. The inventive idea would be the way of achieving the goal. In yet other cases, many people may have a general idea of how they might achieve a goal but do not know how to solve a particular problem, which stands in their way. If someone devises a way of solving the problem, his inventive step will be that solution, but not the goal itself or the general method of achieving it.

In a landmark case, *Conor Medsystems v. Angiotech*, on July 2008 [(2008) UKHL 49], the House of Lords reversed judgments of the Patents Court in 2006 and the Court of Appeal early in 2007 that had both found the U.K.

designation of Angiotech's patent for a stent coated with the drug taxol to be obvious and thus invalid for lacking inventive step. It also raised and made observations on an "obvious to try" argument generally used to arrive at or oppose decisions on inventive step.

The Angiotech patent [EP 0 706 376] was directed to providing a solution to the problem of restenosis, a condition encountered with ordinary stents, where the injury caused to the inner layer of an artery by their insertion could produce an exaggerated healing response that restored the original constriction in the artery that the stent was meant to treat. There was teaching in the specification, based on a particular assay, indicating a likely benefit of using taxol to prevent or treat restenosis, and so it passed the threshold test of making the invention plausible. Having done so, there was no reason why the question of obviousness should be subject to a different test according to the amount of evidence the patentee presented to justify a conclusion that his patent would work. Here the claim in issue was to a stent coated with taxol, the novelty of which was unchallenged. The inventiveness lay not in discovering how to make it but in the claim that such a product would have a particular property, namely, the prevention or treatment of restenosis. Thus, the only relevant question was whether it was obvious to use a taxol-coated stent for this purpose. The test that the lower courts should have applied, but did not, was whether it could be shown based on the prior art "that the skilled person would have an expectation of success sufficient to induce him to incorporate taxol in a drug eluting stent." From what the trial judge at first instance had said in his judgment, it was apparent that he would have answered this particular question in the negative, an answer with which the House of Lords also agreed based on the evidence before the trial judge. Thus, they reinstated the patent.

Lord Hoffmann held that obviousness should be determined by reference to the patent claim "and not to some vague paraphrase based upon the extent of his disclosure in the description."

2.3.5 Australia

As per the Australian Patents Act, an invention is taken to involve an inventive step when compared to the prior art base unless the invention would have been obvious to a person skilled in the relevant art in light of the common general knowledge as it existed in the patent area before the priority date of the relevant claim, whether that knowledge is considered separately or together with either of the kinds of information that is made publicly available in a single document or through doing a single act and being information that the skilled person before the priority date of the relevant claim can be reasonably expected to have ascertained, understood, and regarded as relevant to work in the relevant art in the patent area. Therefore, there is a relevance filter. Another point to consider is whether it was obvious to try and therefore unpatentable if the person versed in the art would assess the likelihood of success under the available facts and circumstances.

The High Court of Australia decision in *Lockwood Security Products Pty Ltd v. Doric Products Pty Ltd* (2007) 235 ALR 202 ("Lockwood") and the U.S. Supreme Court decision in *KSR International Co. v. Teleflex Inc.* (April 30, 2007) ("KSR") throw further light on matters related to inventive step.

In Lockwood, the Australian High Court upheld an appeal by Lockwood, the patentee of an invention for a lock device, finding that the invention was sufficiently inventive, notwithstanding that the problem it sought to solve was well known before the priority date, and that the solution was, at least in hindsight, relatively simple.

The High Court, while considering the question of whether an invention involves an inventive step, observed that the patentee need only show that there was a "scintilla of invention" or that some barrier was crossed or some difficulty was overcome; the "problem solution" approach to the question (wherein a technical problem over the closest prior art information is initially identified and then a consideration is made as to whether the claimed solution, that is, the invention would have been obvious to the skilled worker), favored in Europe, was overly harsh, especially in relation to inventions involving combinations of known components; and secondary evidence of inventiveness should not be lightly ignored. Such evidence includes commercial success, satisfying a long-felt want or need in the relevant art, the failure of others to find a solution to the problem at hand, and subsequent copying of the invention by others. The skilled worker ought to be regarded as a diligent searcher, who would be expected to perform a reasonable search for a solution to a problem at hand rather than a broad search in, for example, divergent or unfamiliar arts (i.e., which might enable the identification of a solution from combinations of features from divergent sources). The High Court also considered that the diligent searcher would not be expected to find every relevant, publicly available document.

2.3.6 Japan

Article 29 (2) of the Japanese Patent Act states that a patent cannot be granted for an invention when that invention could have been easily made, prior to the filing date of the patent application, by a person with ordinary skill in the art to which the invention pertains. The prior art that can be used to make this determination is that described in Article 29 (1)—inventions that have been publicly known or worked anywhere prior to the application for patent, and inventions that have been described in a printed publication anywhere prior to the application for patent.

2.3.7 India

The Indian Patents (Amendment) Act 2005 in Section 2(1)(j) and 2(1)(ja), respectively, states invention means a new product or process involving an inventive step and capable of industrial application. Inventive step means a feature of an invention that involves technical advance as compared to the existing knowledge or having economic significance or both and that makes the invention not obvious to a person skilled in the art. It is interesting to note that the earlier Indian Patents Act (1970) with its amendments in 1999 and 2002 did not have an explicit definition of inventive step.

Further, the Indian Patents (Amendment) Act 2005 in Section 2(1)(l) defines new inventions to mean any invention or technology that has not been anticipated by publication in any document or used in the country or elsewhere in the world before the date of filing of patent application with complete specification, that is, the subject matter has not fallen in public domain or it does not form part of the state of the art. Section 3 of the Indian Patents Act lists what are not inventions despite the inclusion of Section 2(1)(j) defining what an inventive step is, thereby opening an international debate on whether some of the subsections of Section 3 are either redundant or questionable. In the absence of explicit examination guidelines by the Indian Patent Office on how to arrive at the inventive step, the concept is expected to be tested by the judiciary in due course.

2.4 Capable (Susceptible) of Industrial Application (Utility)

An invention is patentable if it is capable (susceptible) of being applicable to any type of industry including agriculture. As has been recognized, the patent system is meant to promote industrial progress and innovation in society facilitated by inventions that have practical applications. In other words, the invention must lead to a useful result. Although this aspect was almost taken for granted in any patent application, issues of credible utility have cropped up in the recent past with emerging fields such as biotechnology and nanotechnology where the intelligent application of the evolving science creates several promising options with possible utility. The questions being addressed are whether these projected promises are specific, substantial, and credible to qualify the benchmark "capable" (susceptible) of industrial application or utility. It ought to be appreciated that this criterion for patentability does not consider whether the invention is capable of being commercialized. It only questions whether the result is useful and is capable of credible industrial application. *In re Ziegler* illustrates how uncertainty can create practical-utility problems. *Ziegler* involved the discovery of polypropylene, but the applicant "disclosed only that solid granules of polypropylene could be pressed into a flexible film…and that the polypropylene was 'plastic-like.'"[9] The court rejected the patent on utility grounds because the application "did not assert any practical use for the polypropylene or its film,"[10] and the mere contention that polypropylene was "plastic like" was insufficient.[10]

Nanotechnology inventions could face serious patentability issues on grounds of uncertainty due to (1) interdisciplinary problems created by the breadth of the nanotechnology field; (2) inoperability problems created by the impossible or inoperable nature of some inventions; (3) practical-utility problems created by the uncertain uses of some inventions; and (4) upstream-research problems created by patents at the research stage of development.[11] Such factors put additional burden on the inventors to provide enabling disclosures with appropriately designed experiments to prove and demonstrate the inventive step and its practical usefulness to overcome any objections on grounds of credible utility.

2.5 Anatomy of a Patent

A patent specification is a structured disclosure of an invention for the purposes of applying for a patent. The general contents of the specification are

- Title
- Name, address, and nationality of inventor and assignee (if any)
- Field of invention
- Background of invention
- Objects of invention
- Detailed description of invention
- Best examples to illustrate the working of the invention
- Claims
- Tables and figures
- Abstract

A patent specification begins with a title that is related to the invention and the supporting subject matter disclosed in the specification. The "field of invention" describes the broad field of the technology and invention being described in the specification and helps to outline the scope of the invention and its purpose. The "background of the invention" provides an introduction to the invention in the context of the relevant prior art and describes the state of art with regard to the problem being addressed and the possible solution gaps, shortcomings in the prior art, and unmet needs that exist at that point in time.

"Summary of invention" or "objects of invention" further defines the contours of the problem and provides the dimensionalities of the problem being addressed around the problem epicenter with a variety of approaches to tackle the defined problem. These are generally done through multiple statements by way of the objects of invention.

In the description, various features of the invention are described completely, explicitly, and clearly. The best method of performing the invention known to the applicant is provided. If the invention relates to a product that is a system, device, or apparatus, then the same is described in the text and illustrated with the help of the drawings. In case the invention relates to a process, then the process steps are described detailing the process parameters.

The invention is described with reference to schematics and drawings wherever necessary. If mathematical formulae are necessary to describe the invention, they are included as appropriate. The invention is then illustrated with the best examples so that the invention is described sufficiently so a person of ordinary skill in the art using his or her common general knowledge is able to reproduce the invention without undue experimentation and use of his or her inventive ability.[2]

In summary, the patent specification is expected to

- Set forth in the description the full and complete details of making and using the invention[12]
- Provide enabling disclosures to teach anyone skilled in the art how to make and use the full scope of the invention without undue experimentation[13]
- Illustrate the invention with the best mode of carrying out the claimed invention known at the time of filing for a patent[14]

The description need not disclose the know-how. It, is therefore, necessary to appreciate the invention in its completeness to decide on what may satisfy the requirements of enabling disclosure.

The claims are so drafted to define the techno-legal contours and the specific areas of the invention for which a protection (monopoly) is being sought.

Generally, there will be one or more claims that are "independent" directed to the essential features of the invention. An independent claim explicitly specifies all of the essential features needed to define the invention. Features that are add-ons to the essential features may be claimed as a set of dependent claims. Dependent claims may be grouped with the independent claims (or within themselves) to include all the features of the independent claim and characterized by additional nonessential features and even the minute aspects and optional features. It is through such a web of claims that the invention is protected in a patent specification.

In most jurisdictions, the claims must *relate to one invention only or to a group of inventions so linked as to form a single general inventive concept.* In case of a group of inventions linked to a single inventive concept, there may be a number of independent claims required. This aspect is known as *unity of the invention.* If there are a set of inventions that do not fit within the concept of *unity of invention,* then those inventions are to be claimed as another patent or as a set of other patents. In the field of nanotechnology, which spans across disciplines and integrates diverse aspects of material properties, processes, functions, constructions, applications, software, devices, etc., it is often difficult to preserve the unity of invention. Under such circumstances, claim construction is tricky and a lot of care has to be exercised while describing the invention and claiming all the features for effective protection and enforcement of the claims in infringement proceedings.

The claim structure is as follows:

- Preamble, which mentions whether it is a product, process (method), or composition of matter ("a device ...", "a composition ...")
- Transition between the preamble and the elements is the transition phrase "comprising ..." or "consisting of ..."
- Critical elements/absolutely essential elements necessary (to distinguish prior art) all in one sentence

The core of the claim is the "elements" section, the one or more elements A, B, and C that define precisely what is claimed. The claim must include elements that distinguish it from the prior art. It is recommended that Claim 1 should include only the "essential" elements. If an element is included within a claim, it is per se an "essential" element from the standpoint of patent infringement. Any accused infringing embodiment without "all elements" of the claim is not a literal infringement.[10]

If the word "comprising" is used, then the additions to the list further limit the invention and are covered by the claim. "Comprising" is an open term. For example, if the claim states "the metal oxide gas sensor comprising elements A and B," this means that the claim covers any metal oxide gas sensor with A and B and any other metal oxide gas sensor that has elements A and B and any other components. The transition term "comprising" is synonymous with "including," "containing," or "'characterized by." These terms are inclusive or open-ended and do not exclude additional, unrecited elements or method steps (see, e.g., Ref. 15).

If the phrase "consisting of..." is used, then the claim is closed to further limitations. For example, if the claim states "the metal oxide gas sensor *consisting of* elements A and B," then the claim is not infringed if the accused embodiment has elements A, B, and any other element. The transitional phrase "consisting of" excludes any element, step, or ingredient not specified in the claim.[16]

A middle ground is the "hybrid" transition "consisting *essentially* of." For example, if the claim states "the metal oxide gas sensor *consisting essentially of* elements A and B," then the "consisting essentially of" language is open to include elements A, B, and C if adding C does not materially change the characteristics of the metal oxide gas sensor, but is "closed" to exclude from infringement a metal oxide sensor with elements A, B, and C where adding C does change the characteristics of the metal oxide sensor. The transitional phrase "consisting essentially of" limits the scope of a claim to the specified materials or steps "and those that do not materially affect the basic and novel characteristic(s)" of the claimed invention.[14] Critical to a properly broad interpretation of the claims is a complementarily drafted specification. The claims must be read in light of the specification, the single best guide to the meaning of a disputed term.[17]

A special definition of each critical term in a "definitions" section in the summary of the invention is warranted without which the usage of the term may be considered to be ambiguous. Putting a term in quotation marks followed by a definition of that term is the safest way to ensure that the term will be interpreted in the manner intended by the applicant. Thus, the specification may reveal a special definition given to a claim term by the patentee that differs from the meaning it would otherwise possess. In such cases, the inventor's lexicography governs.[18]

It is important to use specific defined terminology in the patent claim instead of using broad terms such as "nanoparticles" or "nanostructures." In some cases, the specific types of nanostructure or nanoparticle (e.g., quantum dots, nanotubes, nanowires, nanocups, nanocones, nanoliposomes, nanoshells, nanocrystals, etc.) are ambiguously described in the "written description" section of the patent. The introduction and use of special or nonstandard terms in the claims comes with the obvious risk that a court may construe such terms with a meaning that was unintended by the applicant. Unless the special claim terms are clearly defined in the specification, a court may apply an ordinary meaning such as the dictionary definition of the term. This can adversely affect enforcement of patent rights.[2]

The Federal Circuit Court's *Phillips* decision places a premium on the careful drafting of the specification because terms also may be defined (and limited) unwittingly by implication. Given the court's renewed reliance on claim differentiation, it is important as well to claim nonessential limitations in dependent claims to show the breadth of the independent claims. Interested readers can read the judgment.[19]

The use of the wrong phrases could be fatal to patent protection. For example, the *Chef America* case involved a baking process where dough was flash heated at near incineration temperatures to impart special properties to the final bakery product. However, mistakenly it was claimed that "heating the dough to a temperature in the range of about 400°F to 850°F for a period of time ranging from about 10 seconds to 5 minutes." According to the literal meaning of the claim, it is required that the bakery product reach the near incineration temperature—heating "to" near incineration temperatures—instead of flash heating, that is, "at" the high temperature. The use of "to" instead of "at" makes the claim nonsensical since the end product would be incinerated, inedible carbon instead of the desired tasty bread product.[14]

Following are some illustrations of "nanotechnology" claims construction.

2.5.1 Example 1

The claim 1 of U.S. Patent No. 7068898 entitled "Nanocomposites" issued to Nanosys says the following:

> A composite material, comprising: a matrix; and *one or more nanostructures*, the one or more nanostructures each comprising a core and at least one shell, the core comprising a first semiconducting material having a conduction band and a valence band, the shell comprising a second semiconducting material having a conduction band and a valence band, and the first and second materials having a type II band offset.

Dependent Claim 6 of this patent further narrows the nanostructure limitation of Claim 1 by reciting specific classes of nanostructures:

> A composite material as in claim 1, wherein the one or more nanostructures comprise one or more of: *nanocrystals, nanowires, branched nanowires, or nanotetrapods.*

However, clarification as to the meaning of the "nanostructure" term in Claim 1 is available by referring to a "Definitions" section, which the patentees included in the specification and is as follows (emphasis added):

> A *"nanostructure" is a structure having at least one region or characteristic dimension with a dimension of less than about 500 nm*, e.g., less than about 200 nm, less than about 100 nm, less than about 50 nm, or even less than about 20 nm. Typically, the region or characteristic dimension will be along the smallest axis of the structure. *Examples of such structures include nanowires, nanorods, nanotubes, branched nanowires, nanotetrapods, tripods, bipods, nanocrystals, nanodots, quantum dots, nanoparticles, and the like.* Nanostructures can be substantially homogeneous in material properties, or in certain embodiments can be heterogeneous (e.g., heterostructures). The nanostructures can be fabricated from essentially any convenient material or materials. The nanostructures can comprise "pure" materials, substantially pure materials, doped materials and the like, and can include insulators, conductors, and semiconductors. A nanostructure can optionally comprise one or more surface ligands (e.g., surfactants).

> A "nanoparticle" is any nanostructure having an aspect ratio less than about 1.5. Nanoparticles can be of any shape, and include, for example, nanocrystals, substantially spherical particles (having an aspect ratio of about 0.9 to about 1.2), and irregularly shaped particles. Nanoparticles can be amorphous, crystalline, partially crystalline, polycrystalline, or otherwise. Nanoparticles can be substantially homogeneous in material properties, or in certain embodiments can be heterogeneous (e.g., heterostructures). The nanoparticles can be fabricated from essentially any convenient material or materials. The nanoparticles can comprise "pure" materials, substantially pure materials, doped materials and the like, and can include insulators, conductors, and semiconductors.[2]

2.5.2 Example 2

European Patent application EP 1609 826 issued to Samsung Electronics Ltd. relates to antimicrobial paint containing nano silver particles and a coating method using the same.

Following is the technical problem addressed in this invention:

- Discoloring of paints containing nano silver particles due to ambient temperature.
- Oxidation of silver to silver oxide due to exposure to ultraviolet (UV) rays resulting in yellowing of the paint leading to difficulty in chemically bonding with other objects.
- Failure of nano silver particles contained in paint to maintain the antibacterial properties.

Claim 1 states the following:

> Antibacterial paint comprising nano silver particles, wherein the antibacterial paint is ultraviolet (UV)-curable paint for top-coating of an injection preform of a portable terminal and contains 25 to 35 ppm of nano silver particles having a diameter of 3 to 7 nm

The dependent Claim 2 further narrows the limitation of Claim 1 by reciting specific paint additives in terms of use of nonyellowed, polyurethane-based nano polymer as a photo-polymerized polymer (so as to minimize yellowing phenomena even under strong UV rays).

Independent Claim 3 provides a method for manufacturing antibacterial paint using nano silver particles. It states the following:

> 3. A method for manufacturing antibacterial paint containing nano silver particles comprising the steps of: dispersing nano silver particles into ethanol with a predetermined concentrate to obtain a nano silver-ethanol dispersion solution, which is gradually injected into an isobutanol solution and is agitated continuously for a predetermined period of time to prepare a master batch containing a predetermined concentration of nano silver particles; and injecting the master batch into a UV clear paint and agitating with a high speed for a predetermined period of time to obtain antibacterial UV clear paint containing a predetermined concentration of nano silver particles.

The dependent Claims 4 to 9 (on Claim 3) provide process parameters (and narrow the limitation of Claim 3) in terms of concentration ranges, agitation speed, and time as well as nano particle size range selection.

Independent Claim 10 provides the coating method using this antibacterial paint. It reads as follows:

> 10. A coating method using an antibacterial paint containing nano silver particles comprising the steps of: coating a surface of an injection

preform with a primer covered with color paint; performing a primary drying of the surface of the primer with infrared (IR) rays; coating an upper portion of the printer with a topcoat using UV clear paint prepared by injecting a master batch; and performing a secondary drying of the surface of the topcoat with UV rays.

The dependent Claims 11 to 15 (on Claim 10) provide (and narrow the limitation of Claim 10) in terms of use of type of dryer, temperature and time for drying, use of secondary drying, its temperature, and time.

2.5.3 Example 3

United States Patent 7553760 issued to International Business Machines Corporation (IBM) relates to sub-lithographic nano interconnect structures, and methods for forming the same.

The following technical problems are addressed in this invention:

- Need for scaling up interconnect wiring with scaling up of semiconductor device technology
- The diameters of the openings of the interconnect wiring structures to below the resolutions of the lithographic tools
- Need for sub-lithographic (i.e., below the resolutions of the lithographic tools) feature patterning of interconnect structures for the 64 nm and 32 nm node generations

Claim 1 states the following:

A method of forming an interconnect structure comprising: providing a structure including a patterned hard mask having a lithographically defined opening having a width from about 60 to about 120 nm located above a dielectric material; forming at least one self-assembling block copolymer embedded within said lithographically defined opening, said forming at least one self-assembling block copolymer including applying a layer of a block copolymer comprising polystyrene-block-polymethylmethacrylate (PS-b-PMMA) having a PS:PMMA weight ratio ranging from about 20:80 to about 80:20 over the patterned hard mask, wherein the block copolymer comprises at least a first polymeric component and second polymeric block component that are immiscible with each other and annealing the block copolymer to form a single unit polymer block inside said lithographically defined opening of said patterned hard mask, wherein said single unit polymer block comprises the second polymeric block component comprising an ordered array of cylinders that is aligned vertical to a surface of said dielectric material and is embedded in a polymeric matrix that comprises the first polymeric block component, said cylinders having sub-lithographic widths from about 10 to about 40 nm; selectively removing one of said first or second polymeric components relative to the other component to form sub-lithographic openings

> in said at least one self-assembling block copolymer; transferring said sub-lithographic openings into said dielectric material; and filling said sub-lithographic openings with a conductive material.

The term *sub-lithographic* is used in the claim as well as the title of the invention. Note that the clarification as to the meaning of the term *sub-lithographic* is provided in the specification as "The terms 'nano-scale' and 'sub-lithographic' are used interchangeably throughout the instant application to denote patterns or openings whose widths are below 60 nm" (page 20, column 2, line 2 of the specification).

It is to be noted that there is only one claim in this invention. This claim covers the essential aspects of the invention in terms of description of the article including range of dimension, aspect of self-assembling block copolymer, composition of the block polymer supported with ranges of the constituents, processing of the block polymer (in terms of annealing), constructional details of the polymer block (in terms of array of cylinders), and the aspect of sub-lithographic dimensions and further method of creating sub-lithographic openings and filling with conductive material.

2.5.4 Example 4

United States Patent 7884525 entitled "Carbon Nanotube Based Compliant Mechanism" issued to Massachusetts Institute of Technology relates to a nanoscale compliant mechanism including a coupler and a plurality of nanotubes disposed for nanoscale motion relative to a grounded component.

The following needs are addressed in this invention:

- Nanoscale devices such as grippers for nanomanipulation, force-displacement transmissions for nanoscale transducers, and positioners for probing applications
- Nanoscale version of compliant or rigid link-hinge parallel guiding mechanisms (PGM)

The claim 1 states:

> A nanoscale compliant mechanism comprising: a grounded component; a mechanical coupler disposed to move relative to the grounded component; and a plurality of nanotubes having first and second ends, the first ends of the nanotubes being coupled to the grounded component and the second ends of the nanotubes being coupled to the coupler, the nanotubes being compliant relative to the grounded component and the coupler; wherein the coupler is disposed for substantially linear motion relative to the grounded component.

This claim protects essential components of the compliant mechanism such as (1) grounded component, (2) coupler, and (3) nanotubes along with functional relation of the nanotubes, coupler, and grounded components.

The dependent Claims 2 to 6 further narrow the limitation of Claim 1 by reciting nanotube material (carbon nanotube claimed in Claim 2) and type of nanotube (single- or double-walled carbon nanotube as claimed in Claim 3).

The dependent Claim 5 further narrows the "coupling" aspect of Claim 1 by reciting specific types of coupling in terms of "bonding" of the nanotube with the ground component and the coupler. Claim 6 further provides selection of bond from materials/types.

The independent Claim 7 states:

> A nano-scale compliant mechanism comprising: a grounded component; a mechanical coupler disposed to move relative to the grounded component; and a plurality of nanotubes having first and second ends, the first ends of the nanotubes being coupled to the grounded component and the second ends of the nanotubes being coupled to the coupler, the nanotubes being compliant relative to the grounded component and the coupler; wherein displacement of the coupler relative to the grounded component induces a bending strain in the nanotubes.

Claim 1 covered the *linear motion* aspect of the coupler relative to the grounded component. The independent Claim 7 covers the aspect of induction of *bending strain* in the nanotubes.

The dependent Claims 8 to 12 further narrow the limitation of independent Claim 7 related to bending strain values, coupler displacement with respect to nanotube length, cyclic loading/unloading aspect, and mechanical behavior of the coupler.

The independent Claim 13 states (emphasis added):

> A nano-scale parallel-guided mechanism comprising: a grounded component; a mechanical coupler; and a plurality of carbon nanotubes each having first and second ends and a *longitudinal axis, the longitudinal axes of the nanotubes being substantially parallel with one another,* the first end of each of the nanotubes being coupled to the grounded component, the second end of each of the nanotubes being coupled to the coupler, the nanotubes being compliant relative to the grounded component and the coupler.

The aspect of disposition of nanotubes parallel to each other is introduced in this independent claim. Claim 1 stated only the coupling aspect of the nanotubes and does not state disposition of nanotubes with each other.

Claims 14 to 21 are dependent claims (on Claim 13) and claim types of nanotubes, selection of nanotubes, loading and unloading cycle aspect, mechanical behavior, and bending strain aspect specific to the variant cited in independent Claim 13.

Further, the independent Claim 22 cites (emphasis added):

> A nano-scale compliant mechanism comprising: a grounded component; a mechanical coupler disposed to move relative to the grounded component; and a plurality of nanotubes having first and second ends,

the first ends being coupled to the grounded component and the second ends being coupled to the coupler; *wherein displacement of the coupler relative to the grounded component induces a bending strain in the nanotubes, said displacement exhibiting first and second regions of mechanical behavior:* the first region including relatively low bending strain; and the second region including relatively high bending strain, the second region being predominantly governed by compliant, hinge-like bending of at least one kink in the nanotubes.

This independent Claim 22 covers the combination of aspects of both bending strain as well as mechanical behavior that are separately claimed in independent claims. The dependent Claims 23 to 28 claim types of nanotubes, selection of nanotubes, loading and unloading cycle aspect, mechanical behavior, and bending strain aspect specific to the variant cited in independent Claim 22.

2.5.5 Example 5

European Patent Application EP2351702 issued to Korea Institute of Machinery & Materials relates to apparatus and a method for manufacturing a quantum dot. The following technical problems are addressed in this invention:

- Apparatus and method for mass production of quantum dots
- Mass production of quantum dots with uniform particle diameters at high yield

Claim 1 cites the following:

An apparatus of producing quantum dots, comprising: at least one pump to inject a plurality of precursor solutions in which different kinds of precursors are dissolved at a predetermined flow rate; a micro mixer in which a plurality of precursor solutions are mixed in such a manner that a plurality of paths diverge from each of a plurality of input ports to which the precursor solutions are respectively supplied, the diverging paths joining with other paths diverging from the other one of the input ports, and then the joined paths are collected into an output port; and a heating furnace to pass the precursor mixture solution through to create and grow quantum dot nucleuses, thus producing quantum dots.

The essential components of apparatus such as pump, micro mixer, and heating furnace are claimed in Claim 1.

The dependent Claims 2 to 5 claim the components such as mixer heating unit, heating furnace, mixer heating unit functionality, and the use of a cooling unit. It can be noted that the provisions for heating and variants are claimed in this set of dependent claims.

The independent Claim 7 reads as follows:

> A method of producing quantum dots, comprising: pumping a plurality of precursor solutions in which different kinds of precursors are dissolved to supply each precursor solution at a predetermined flow rate and at a constant speed; mixing the plurality of precursor solutions by diverging each precursor solution into a plurality of micro streams, joining each diverging micro stream to a micro stream diverging from another precursor solution and then collecting the joining streams; and heating the precursor mixture solution to a predetermined temperature to create and grow quantum dot nucleuses, thus producing quantum dots.

The independent Claim 6 provides a method of producing quantum dots. The aspect of micro stream divergence and joining of the different diverging micro streams is claimed as a part of the method. However, Claim 1 covered the micro mixer constructional aspect with respect to input and output ports rather than claiming formation and further mixing of micro streams.

The dependent Claims 7 to 9 are the variants of the method in terms of preparation of precursor solutions, temperature aspect, and cooling of the heated precursor solution.

2.5.6 Example 6

United States Patent 7824626 issued to Applied Nanotech Holdings, Inc. relates to an air handler and purifier. The following technical problems are addressed in this invention:

- Eliminate indoor pollutants, and chemical and biological agents within an HVAC system.
- Neutralize biological threats.

Claim 1 cites the following:

> Filter material for gas and liquid comprising: a support layer; a layer of binder on the support layer in which ionic silver and/or metallic silver particles are incorporated; and a layer comprising particles of photocatalyst deposited on the layer of binder, wherein loading of ionic silver varies from 0.1 to 10 grams per square meter of the filter material, and loading of photocatalyst particles varies from 1 to 100 grams per square meter of the filter material, and wherein the layer comprising particles of photocatalyst is a topmost layer adjacent to an ongoing flow of the gas or the liquid.

The essential constructional components of the filter, such as (1) support layer, (2) a layer of binder on the support layer incorporated with ionic silver or metallic silver particles, and (3) a layer comprising particles of photocatalyst and their relative disposition with respect to each other, is claimed in

Claim 1. Further, it is to be noted that Claim 1 characterizes the density range of the silver in terms of grams per meter square.

The dependent claims (on Claim 1) from 2 to 9 claim variants of the support layer in terms of diverse material selection, form of ionic silver, binder, and photo catalyst material.

Claim 10 is an independent claim that addresses the aspect of irradiation of the ionic silver or metallic silver particles by UV radiation.

Claims 11 to 16 are dependent on Claim 10. The variants of support layer in terms of diverse material selection, ionic silver variant, binder, and photo catalyst material for the filter material of Claim 10.

The independent Claim 17 claims the entire fluid treatment apparatus. The claim structure of this invention is important to note. Claim 1, Claim 10, and the respective dependent claims protect the aspect of filter media. However, Claim 17 protects the entire apparatus provided with inlet, outlet, disposition of filter media, and source of UV light.

2.5.7 Example 7

United States Patent 8062697 issued to Applied Nanotech Holdings, Inc. relates to ink jet application for carbon nanotubes. The following technical problem is addressed in this invention:

- Inexpensive way of applying the CNT material onto suitable substrate materials at low temperature and aligning these materials using methods that are suitable for large-scale manufacturing.

Claim 1 cites the following:

> A method for making a field emission cathode comprising the step of selectively depositing a field emitter mixture onto a substrate using an ink jet dispenser, wherein the field emitter mixture further comprises carbon nanotubes and other nanoparticles comprising other forms of carbon.

Claim 1 concerns essential aspects of the method, namely deposition of the mixture, use of ink jet dispenser, and use of carbon nanotubes and other forms of carbon.

Claims 2 to 6 are dependent claims on Claim 1. The variants of other forms of carbon are claimed in these claims.

Independent Claim 7, in addition to the method claimed in Claim 1, covers particular characteristics of the nanoparticles in terms of insulating or wide-band gap particles.

Claim 8 is dependent on Claim 8, which covers selection of the insulating or wide-band gap particles.

Claims 9 and 10 are independent claims. Each of the claims, in addition to the aspects claimed in Claim 1, covers characteristics of the nanoparticles such as semiconductor nanoparticles, and selection of nanoparticles from the group consisting of clays, clay particles, and organo-clays. Claim 11 is another independent claim that covers method and equipment used in terms of ball mills for grinding carbon nanotubes.

2.6 Patenting Systems: An Overview

Nanotechnology as a field has become a favorite ground for strategic planning of patents to create impregnable patent estates to enable the patent to secure the national right, in the country granting the patent, and to exclude others from commercially utilizing the claimed invention for a limited period of time without the consent of the patent holder. In return, the patent holder has to disclose the invention in a patent specification to enable a person skilled in the art to reproduce the invention without undue experimentation. Thus, a patent is a contract between the government/state and the inventor wherein the state gives the inventor a monopoly to the invention in exchange for a "full public disclosure" of the invention in a patent specification. The period for which the monopoly is granted is 20 years from the date of filing the patent application in that country. However, it must be appreciated that the monopoly is not automatic, as the patent application has to be processed through a formal examination process in the patent office. A general awareness of the process steps and effective utilization of the services of a profession patent attorney/agent are always helpful.

It is essential to apply for a patent in the country of interest. Every country has specific requirements and processes for patent applications. It is to be appreciated that patents are national rights and to obtain rights in various countries, it is necessary to apply for patents and follow the processes as required by the respective patent laws in each country.

In all countries (except until recently in the United States), the applicant can be the inventor or an assignee. In the United States until September 2011, patent applicants could only be the inventor. Even if the inventors assigned their invention, the assignee could not become a patent applicant.

All patent offices around the world have been following a system termed "first-to-file" wherein the right to the grant of a patent for a given invention lies with the first person to file a patent application for protection of that invention, regardless of the date of actual invention.

Until recently, the United States was following a system that was termed "first to invent." Under this system when an inventor conceives of an invention and *diligently* reduces the invention to practice, the inventor's date of invention is the date of conception. Accordingly, the person first to conceive the invention and diligently reduce it to practice is considered the first inventor and the inventor entitled to a patent, even if another person files a patent application for the same invention before the inventor.

On September 16, 2011, President Obama signed the Leahy–Smith America Invents Act (The AIA or Patent Reform Act of 2011), which brought new legislation to the American patenting system changing the patent system from "first-to-invent" to "first-to-file" and thereby falling in line with the majority of patent systems globally. Additionally, the new legislation permits an inventor to assign another party to apply for a patent on his or her behalf. These changes substantially align the U.S. patent system with the rest of the systems being followed globally.

The general principles of the patent laws and processes in various countries are based on the TRIPS Agreement. A few of the relevant articles of the TRIPS Agreement are reproduced here for the reader to appreciate the obligations of the member states of the WTO with regard to their respective intellectual property laws. The detailed text of the TRIPS Agreement can be accessed at the WTO website http://www.wto.org/english/docs_e/legal_e/27-trips.01_e.htm

Thus, there are

- National patenting systems
- International conventions such as
    - The Paris Convention under which the first patent filing date in one country, known as the priority date, is recognized in all the other countries of the convention provided that further applications are made in the member states of the Paris Convention within 12 months of the priority date. Patents are not granted under this convention. It only facilitates the filing of patents in various member states within a window of 12 months from the priority date of the patent application.
    - Patent Cooperation Treaty (PCT) is a multilateral treaty signed in Washington on June 19, 1970, amended on September 28, 1979, and modified on February 3, 1984, and on October 3, 2001. The PCT is administered by the World Intellectual Property Organization (WIPO), between more than 140 Paris Convention countries. The states party to this Treaty, which are also called "the contracting states," constitute a union for cooperation in the filing, searching, and examination of applications for the protection of inventions, and for rendering special technical services. This union is known as the International Patent Cooperation Union (Article 1: Establishment of a Union, Patent Cooperation Treaty). Applications for the protection of inventions in any of the contracting states may be filed as international applications under this Treaty (Article 3: International Application, Patent Cooperation Treaty). It is important to note that a national or resident of one of the PCT contracting states is entitled to file an international patent application.

- PCT makes it possible to seek patent protection for an invention simultaneously in each of a large number of countries by filing an "international" patent application. It establishes a procedure for the *filing* and *processing* of a single application for a patent, which has legal effect in the PCT contracting states with an extended period up to 30 months or 31 months (in some contracting states) from the priority date of the patent application. This means that the applicant gets 18 additional months over the Paris Convention (which gives 12 months from the priority date for national phase filings from the priority date) to file the application in the various national jurisdictions that are PCT contracting states.

- Such an application may be filed by anyone who is a national or resident of a contracting state. It may generally be filed with the national patent office of the contracting state of which the applicant is a national or resident or, at the applicant's option, with the International Bureau of WIPO in Geneva. If the applicant is a national or resident of a contracting state that is a party to the European Patent Convention, the Harare Protocol on Patents and Industrial Designs (Harare Protocol), the revised Bangui Agreement Relating to the Creation of an African Intellectual Property Organization, or the Eurasian Patent Convention, the international application may also be filed with the European Patent Office (EPO), the African Regional Industrial Property Organization (ARIPO), the African Intellectual Property Organization (OAPI), or the Eurasian Patent Office (EAPO), respectively.

- It must be appreciated that PCT is only a system that facilitates the filing of patent applications in various contracting states; PCT is not a patent granting system.

Countries such as Argentina, Venezuela, Bangladesh, Saudi Arabia, and Taiwan are not yet PCT contracting states and, therefore, the PCT system does not apply to these countries. *It is important to note that the applicant needs to file the patent application within 12 months from the first filing (priority date) in these countries for their applications to be prosecuted as valid patent applications in these countries.*

- Regional arrangements such as
  - European Patent Convention (EPC) covers 31 contracting states as follows: (AT) Austria, (BE) Belgium, (BG) Republic of Bulgaria, (CH) Switzerland, (CY) Cyprus, (CZ) Czech Republic, (DE) Germany, (DK) Denmark, (EE) Republic of Estonia, (ES) Spain, (FI) Finland, (FR) France, (GB) United Kingdom, (GR) Hellenic Republic, (HU) Hungary, (IE) Ireland, (IS) Iceland,

(IT) Italy, (LI) Liechtenstein, (LT) Lithuania, (LU) Luxembourg, (LV) Latvia, (MC) Monaco, (NL) Netherlands, (PL) Poland, (PT) Portugal, (RO) Republic of Romania, (SE) Sweden, (SI) Slovenia, (SK) Slovak Republic, and (TR) Turkey.

- Further, European patents can be extended to the following extension states: (AL) Albania, (BA) Bosnia and Herzegovina, (HR) Croatia, (MK) the former Yugoslav Republic of Macedonia, and (YU) Serbia and Montenegro. These extension states, while not having signed the EPC, have amended their national laws based on the EPC.

- European patents can be applied for by filing an application either directly at the EPO in The Hague, Munich, or Berlin, or via a local national receiving office, such as the U.K. Patent Office. European patents can also be obtained via a PCT patent application designating the European region. Patent applications filed at the EPO are processed in a single official language, which can be English, French, or German.

- At the end of the patent application procedure, the EPO grants a single European patent covering all designated states. To continue the rights for their maximum 20-year term from first filing date, the patent must be "validated" in each of the states for which rights are to be continued. Renewal fees must be paid in each national state for as long as the individual national patents are to remain in force.

- ARIPO was the creation of the joint efforts of the United Nations Economic Commission for Africa (UNECA) and WIPO. This culminated in the Lusaka Agreement in December 1976 and came into force on February 15, 1978. ARIPO covers 17 countries of English-speaking Africa (Anglophone Africa). ARIPO office administers patents under the Harare Protocol with its headquarters in Harare, Zimbabwe. At present, ARIPO is composed of the following member states, namely, Botswana, Gambia, Ghana, Kenya, Lesotho, Liberia (from March 24, 2010), Malawi, Mozambique, Namibia, Rwanda (from September 24, 2011), Sierra Leone, Sudan, Swaziland, Tanzania, Uganda, Zambia, and Zimbabwe.

- ARIPO is a member of the Paris Convention for the Protection of Industrial Property and the Patent Cooperation Treaty (PCT).

- The ARIPO regional system provides the option of filing applications via the member state route and the ARIPO office route by filing a single application in English either at one of the contracting state offices or directly at the ARIPO office. Each application may be designated to any or all members of the ARIPO states where the applicant wishes to have his or her patent protected. Once a patent has been granted, subsequent actions are to be carried through a centralized processing mechanism.

- Applications under the Paris Convention can be filed within 12 months of the priority date and PCT applications may be filed in ARIPO in the national phase within 31 months of the priority date. It is to be noted that a PCT patent application will not extend to those member states that are not PCT members.

- Organisation Africaine de la Propriete Industrielle (OAPI) is the creation of the Bangui agreement, done at Bangui, Republic of Central Africa on March 2, 1977 and revised on February 24, 1999. The countries under this convention are the French-speaking countries (Francophone Africa), namely Benin, Burkina Faso, Cameroon, Central African Republic, Chad, Congo, Equatorial Guinea, Gabon, Guinea, Guinea Bissau, Ivory Coast, Mali, Mauritania, Niger, Senegal, and Togo.

- OAPI is a member of the International Convention and of the PCT. Applications under the Paris Convention can be filed within 12 months of the priority date and PCT applications may be filed in OAPI in the national phase within 30 months of the priority date.

- An OAPI application automatically designates all member states. The OAPI patent office is in Yaounde in the Republic of Cameroon where the applications from outside the union countries are filed. Patent applications originating within OAPI may also be directly filed with the OAPI office in Yaounde. However, certain member states require "indirect filing," which means that the applications must be filed at their national patent office or transmitted from those to Yaounde. While the application language of OAPI applications used to be exclusively French, this is no longer the case and documentation in either French or English is now generally accepted. In fact, the official application form is now in both languages. It is to be noted that a PCT patent application will not extend to those member states that are not PCT members.

- Eurasian Patent Convention (EAPC) is an international patent law treaty instituting the EAPO. It was signed on September 9, 1994 in Moscow, Russia. The present patent regulations under the Patent Convention with amendments and additions were adopted by the Administrative Council of the Eurasian Patent Organization at its 17th (12th ordinary) session on November 14–18, 2005 and came into force on November 18, 2005. The following states are included in the EAPC: (AZ) Azerbaijan, (AM)

Armenia, (BY) Belarus, (GE) Georgia, (KZ) Kazakstan, (KG) the Kyrgyz Republic, (MD) Republic of Moldova, (RU) Russian Federation, (TJ) Republic of Tajikistan, and (UA) Ukraine. The Eurasian Patent Office is located in Moscow and the convention is administered by WIPO. Persons from any United Nations state, bound by the Paris Convention for the Protection of Industrial Property and the PCT, can apply for patent protection in all countries bound by the treaty by filing a Eurasian patent application in the Russian language. Alternatively, Eurasian patents can be applied for via the PCT route.

In all the patent systems, after the patent application is filed, it undergoes an examination at the appropriate patent office wherein an examiner accesses it for patentability based on the national laws that include benchmarks for novelty, inventive step, capable of industrial application/industrial applicability/utility, and further taking into consideration the exclusions to patentability prescribed by the law. The examiner also assesses the enabling disclosures. An examination report is generally sent to the applicant, who has to satisfy the queries raised by the examiner. This is an interactive process between the applicant (or his or her patent attorney/agent) and only on meeting all the queries of the examiner either by argument or by appropriate amendment, or both, the patent may be granted. All patent applications are generally published 18 months from the priority date of the patent application and in some countries, such as in India, a person in the public can initiate pregrant opposition proceedings at the patent office opposing the patent application. In some countries, such as the United States or the European Union, there is no provision for pregrant opposition and the public can oppose the patent only in a postgrant proceeding.

The systems followed in USPTO, EPO, JPO, and the Indian Patent Office are depicted in Figure 2.4 to Figure 2.8.

Whichever route is used, the result will be a "family" of related patent applications for the same invention that will be examined by the patent authorities of the major jurisdictions. Further, as each patent office examines the same invention from its perspective, the claims granted in various countries can vary, which may result in variable protection for the same invention in different jurisdictions.

Apart from the opposition period, it is also possible to challenge the validity of a patent throughout its duration, depending on the laws of the country concerned. The validity of a patent is usually challenged by a defendant during infringement proceedings.

Patents only stay in force if renewal fees are paid at regular intervals based on the patent law in each country, throughout the lifetime of the patent. Failure to pay such renewal fees results in the lapse of the patent, although failure to pay such fees may be rectified and the patent restored if action is taken within a prescribed period.

The Grant Procedure at a Glance

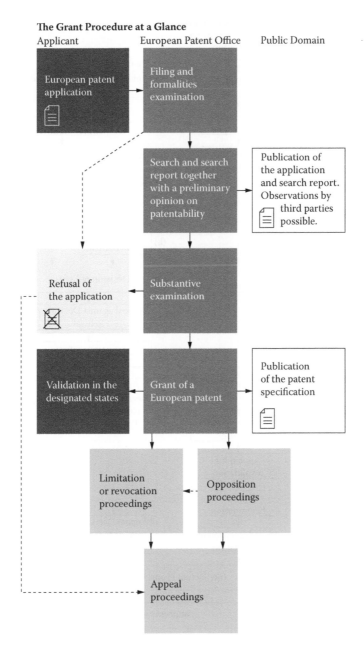

FIGURE 2.4
European patent system (European patents and the grant procedure, www.epo.org).

Procedures for Obtaining a Patent Right

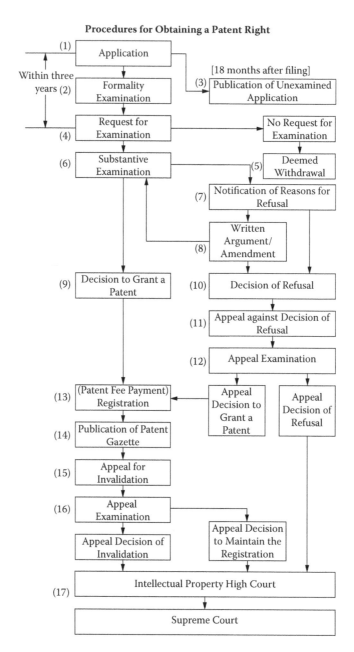

FIGURE 2.5
Japanese patent system (www.jpo.go.jp).

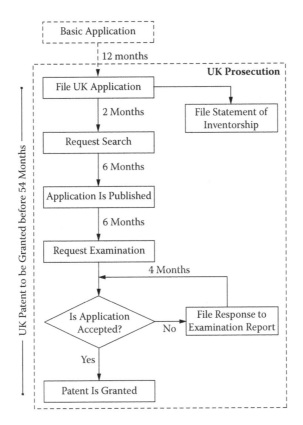

FIGURE 2.6
United Kingdom patent system (http://www.patentretriever.com/GetUKPatentCheaply/GetUKpatentCheaply.html).

2.7 The PCT Process

The various stages in the PCT process are depicted in Figure 2.9.

A PCT international patent application (filed in the prescribed format) has the effect of a national patent application (and certain regional patent applications) in all PCT contracting states. The contents of the PCT international application are PCT request, description of the invention, claims, abstract, and drawings. It is possible to claim priority of the original application. If a PCT application is filed as the first application (priority application), then the date of filing the PCT application is taken as the priority date of the application. It is also possible to file a PCT application based on multiple patent applications filed in one of the contracting states, but the application with the earliest priority date is taken for calculating the 30- or 31-month window for the national phase filings.

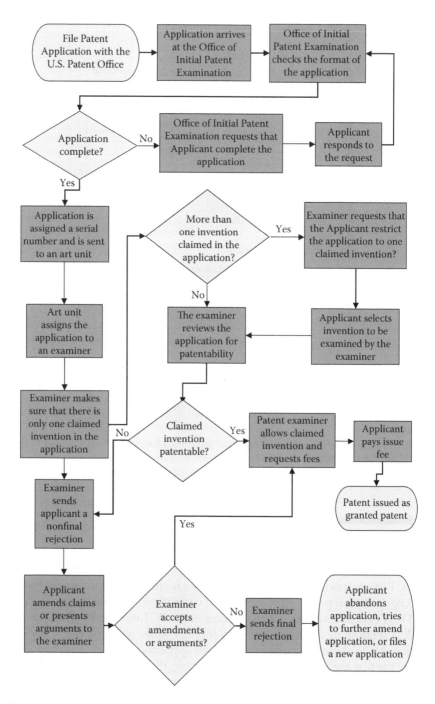

FIGURE 2.7
United States patent system (http://www.inventorbasics.com/patent%20process.htm).

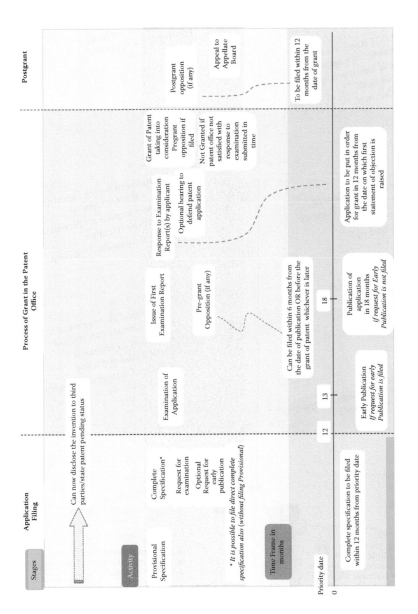

FIGURE 2.8
Indian patent system.

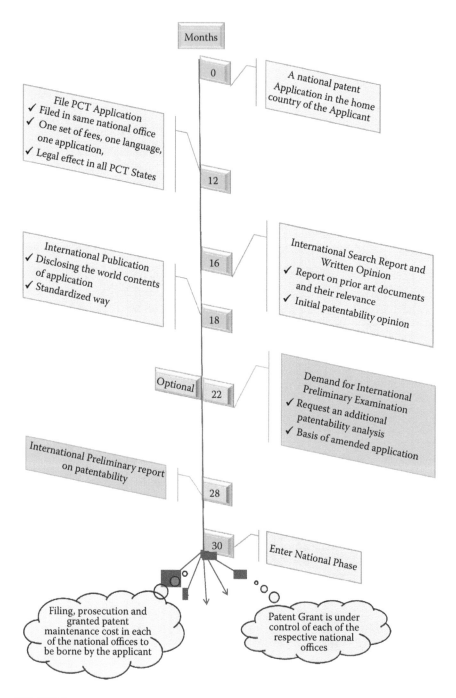

FIGURE 2.9
PCT system.

The international application is then subjected to what is called an "international search." That search is carried out by one of the major patent offices appointed by the PCT Assembly as an International Searching Authority (ISA). The national offices of Australia, Austria, Canada, China, Finland, Japan, the Republic of Korea, the Russian Federation, Spain, Sweden, the United States, and the EPO are appointed as ISAs.

The said search results in an "international search report"; that is, a listing of the citations of such published documents that might affect the patentability of the invention claimed in the international application. At the same time, the ISA prepares a written opinion on patentability.

The international search report and the written opinion are communicated by the ISA to the applicant who may decide to withdraw his or her application, in particular where the said report or opinion makes the granting of patents unlikely.

If the international application is not withdrawn, it is, together with the international search report, published by the International Bureau. The written opinion is not published. The international publication puts the world on notice of the application, which can be an effective means of advertising and looking for potential licensees.

Based on the search report and opinion, the applicant may decide to proceed to the national phase to file the application in various countries within the 30- to 31-month period as permitted by the contracting state. The applicant may amend the application and then file the national phase applications overcoming the objections in the PCT search report and opinion.

Optionally at this stage instead of filing national phase applications, the applicant may file a demand for Chapter II within the PCT system. The applicant submits the application either with amendments or without amendments accompanied by detailed technical explanations and arguments justifying the claims in the patent application and seeking a substantive examination by the ISA as part of the Chapter II proceedings. The ISA then subjects the PCT application to a substantive examination and issues an International Preliminary Report on Patentability (IPRP).

On several occasions, the present authors have experienced that when a Chapter II demand is filed with appropriate technical arguments and amendments, if necessary, justifying the invention in terms of its novelty, inventive step, and industrial applicability vis-a-vis the cited prior art in the international search report, it is possible to get a clean report on patentability of the claims as such or as amended in the IPRP.

Such a situation is of immense use to the applicant as he can then proceed to the national phase with his application with enhanced confidence. The IPRP report is sent by WIPO to the patent offices of the contracting states that request it.

The international preliminary examination report is given due consideration by the national patent offices, but it is not binding on them to accept the conclusions even though the ISA has conducted a substantive examination.

However, a favorable IPRP strengthens the case for an applicant in the national phase proceedings.

2.7.1 Advantages of PCT

The procedure under the PCT has great advantages for the applicant, the patent offices and the general public:

- The applicant has up to 18 months more than he has in a procedure outside the PCT to reflect on the desirability of seeking protection in foreign countries, to appoint local patent agents in each foreign country, to prepare the necessary translations, and to pay the national fees; he is assured that, if his international application is in the form prescribed by the PCT, it cannot be rejected on formal grounds by any designated office during the national phase of the processing of the application; on the basis of the international search report or the written opinion, he can evaluate with reasonable probability the chances of his invention being patented; and the applicant has the possibility during the international preliminary examination to amend the international application to put it in order before processing by the designated offices.

- The search and examination work of patent offices can be considerably reduced or virtually eliminated thanks to the international search report, the written opinion and, where applicable, the international preliminary examination report that accompany the international application.

- Since each international application is published together with an international search report, third parties are in a better position to formulate a well-founded opinion about the patentability of the claimed invention.

- The international publication puts the world on notice of the patent application, which can serve as an effective means of communication with potential licensees.

- Major costs associated with international patent protection can be deferred until one has evaluated all possible options for commercialization of the invention.

References

1. Kroll, H., Exploring the validity of patent applications as an indicator of Chinese competitiveness and market structure, *World Patent Information*, 33, 23–33, 2011.

2. O'Neill, S., Hermann, K., Klein, M., Landes, J., and Bawa, R., Broad claiming in nanotechnology patents: is litigation inevitable? *Nanotechnology Law & Business*, 595, 2007.
3. Nogueira de Sousa Branquinho Nordberg, A.R., Nanotechnology patents in Europe: Patentability exclusions and exceptions, Master's thesis, Stockholm University, 2009, http://www.juridicum.su.se/juruppsatser/2010/ht_2010_Ana_Rita_Nogueira_de_Sousa_Branquinho_Nordberg.pdf.
4. *Verdegaal Bros., Inc. v. Union Oil Co.*, 814 F.2d 628, 631, 2 U.S.P.Q.2d (BNA) 1051, 1053 (Fed Cir. 1987). ("A claim is anticipated only if each and every element as set forth in the claim is found, either expressly or inherently described, in a single prior art reference.")
5. *BASF v. Orica Australia Boards of Appeal of the EPO*, T-0547/99 (January 8, 2002).
6. Watal, A. and Faunce, T.A., Patenting nanotechnology: Exploring the challenges, *WIPO Magazine*, April 2011, http://www.wipo.int/wipo_magazine/en/2011/02/article_0009.html#12.
7. *SmithKline Beecham Biologicals v. Wyeth Holdings Corporation*, Boards of Appeal of the EPO, T-0552/00, October 30, 2003.
8. Ganguli, P., Inventive step—its shades and shadows, Express Pharma, www.expresspharmaonline.com/20090715/management02.shtml.
9. *Ziegler*, 992 F.2d 1197, Fed. Cir. 1993. http://federal-circuits.vlex.com/vid/in-re-karl-ziegler-and-heinz-martin-37539098.
10. Wegner, H.C., Claim Drafting: Unique American Challenges, Foley TACPI.
11. Almeling, D.S., Patenting nanotechnology: Problems with the utility requirement, *Stanford Technology Law Review*, 2004, http://stlr.stanford.edu/pdf/Almeling-PatentingNanotech.pdf.
12. Axford, L.A., Patent drafting considerations for nanotechnology inventions, *Nanotechnology Law & Business*, 3, 305, 2006.
13. Schwaller, M.D. and Goel, G., Getting smaller: What will enablement of nanotechnology require? *Nanotechnology Law & Business*, 3, 145, 2006.
14. Dowd, M.J., et al., Nanotechnology and the best mode, *Nanotechnology Law & Business*, 2, 238, 2005.
15. See, for example, *Mars Inc. v. H.J. Heinz Company Lp*, 377 F.3d 1369, 1376, Fed. Cir. 2004; Harold C. Wegner, Claim Drafting: Unique American Challenges, Foley TACPI.
16. *In re Gray*, 53 F.2d 520, CCPA, 1931; *ex parte Davis*, 80 USPQ 448, 450, Bd. App., 1948; Harold C. Wegner, Claim Drafting: Unique American Challenges, Foley TACPI.
17. *Aero Products Intern., Inc. v. Intex Recreation Corp.*, F.3d, Fed. Cir., 2006; Schall, J. quoting Phillips, 415 F.3d at 1315, quoting *Vitronics Corp. v. Conceptronic, Inc.*, 90 F.3d 1576, 1582, Fed. Cir., 1996.
18. *Cook Biotech Inc. v. Acell, Inc.*, 460 F.3d 1365, 1373, Fed. Cir., 2006; quoting Phillips, 415 F.3d at 1316; citing *CCS Fitness, Inc. v. Brunswick Corp.*, 288 F.3d 1359, 1366, Fed. Cir., 2002; Harold C. Wegner, Claim Drafting: Unique American Challenges, Foley TACP.
19. Federal Circuit Clarifies the Methodology for Performing Claim Construction, *Phillips v. AWH Corp.* Federal Circuit (en banc), July 12, 2005.

3

Looking for Nanotechnology Prior Art

The prolific growth of nanotechnology-related literature in the form of research journals, conference/workshop proceedings, commercial and trade journals, news announcement, promotional and journalistic writings, networking in the Web, and patent publications poses a stiff challenge of acquiring targeted, relevant, and timely information in nanotechnology. Sifting information from the dense information nano-jungle has been difficult especially due to its all-pervasive multidisciplinary nature.

It is well recognized that a significant portion of information related to nanotechnology resides in patent applications and granted patents that does not appear in other publication platforms.

The global patent system has undergone a metamorphic transformation in the last few decades with the growth of electronic storage, retrieval, and communication systems.

It is also recognized that the patent system is one of the most well structured, indexed, and retrievable documentation of technology especially as it softly links science and technology with usefulness facilitating the movement of ideation toward commercialization.

The patent databases can be used to derive technical information including ongoing research and thrust areas in emerging technologies, identify potential threats from competitors, explore potent collaborators for R&D, map opportunities for IP transactions, analyze the information to strategize business opportunities in terms of new product launches, co-marketing of products and services in diverse jurisdictions, and assess options for setting up strategic alliances, mergers, acquisitions, and joint ventures based on technological strengths.

Further, the patent databases can be used to design planned protection of one's own developed technologies both for defensive purposes by ensuring freedom to operate in selected jurisdictions and for offensively blocking competitors. The technical information including ongoing research and thrust areas in emerging technologies is of immense significance as it helps to avoid duplication of efforts, find state of the art solutions to existing problems, generate ideas for R&D, and develop approaches to bypassing existing patents and avoiding the dangers of knowingly infringing patents.

Mapping of the technology landscape with the associated intellectual property grids is increasingly being used by businesses as well as policy makers in their planning exercises. Audit and asset valuation of patent portfolios of various commercial houses, academic and research organizations, and entrepreneurs are being used by financial institutions, venture capital agencies, and banks to arrive at funding decisions.

Effective information search therefore involves strategic planning and search in diverse information sources utilizing a range of tools and methods to establish the appropriate prior art in nanotechnology. It is an absolute imperative that exhaustive prior art search is carried out and the field of interest landscaped prior to embarking on any nanotechnology project.

3.1 International Patent Classification System

The International Patent Classification (IPC) as a result of the Strasbourg Agreement Concerning the International Patent Classification was established on March 24, 1971, and amended on September 28, 1979. It is a hierarchical system in which all fields of technology are divided into eight sections, A to H, which are further subdivided into classes, subclasses, groups, and subgroups. The sections are classified as A: Human necessities; B: Performing operations, transporting; C: Chemistry, metallurgy; D: Textiles, paper; E: Fixed constructions; F: Mechanical engineering, lighting, heating, weapons, blasting engines or pumps; G: Physics; H: Electricity.

Nanotechnology inventions fall into the B82B subclass of IPC that is entitled "Nano-Structures; Manufacture or Treatment Thereof." As stated in the notes that explain this subclass, B82B does not cover chemical or biological structures per se, as these are dealt with elsewhere.[1]

As per Articles 1 and 2 of the Strasbourg Agreement, all contracting parties, which includes all European Patent Office (EPO) member states, are bound to follow the international classification of patents that was established pursuant to the provisions of the European Convention on the International Classification of Patents.[2]

Although all member states follow the IPC, some patent offices have developed their own internal classifications or tagging codes. For example, for tagging nanotechnology-related patents, the U.S. patent office defined subclass 977, the European patent office defined class Y01, and the Japanese patent office created the ZNM class.

The next sections elaborate on the system followed by the three major patent offices in the world.

3.1.1 European Patent Office (EPO)

The European Classification System (ECLA) is an extension of the IPC used by the EPO. In collaboration with the nanotechnology working group (in 2003, the EPO created this group with the aim of confronting challenges posed by nanotechnology), the documentation department of the EPO created a

TABLE 3.1

B82Y Nanotechnology Subclass

Code	Title
B82Y	Nanotechnology
B82Y5/00	Nanobiotechnology or nano-medicine, e.g., protein engineering or drug delivery
B82Y10/00	Nanotechnology for information processing, storage, or transmission, e.g., quantum computing or single electron logic
B82Y20/00	Nanotechnology for interacting, sensing, or actuating, e.g., quantum dots as markers in protein assays or molecular motors
B82Y20/00	Nano-optics, e.g., quantum optics or photonic crystals
B82Y25/00	Nano-magnetism, e.g., magneto-impedance, anisotropic magneto-resitance, giant
	Magneto-resistance or tunneling magneto-resistance
B82Y30/00	Nanotechnology for materials or surface science, e.g., nano-composites
B82Y35/00	Methods or apparatus for measurement or analysis of nano-structures
B82Y40/00	Manufacture or treatment of nano-structures
B82Y99/00	Subject matter not provided for in other groups of this subclass

Source: Nanotechnology and Patents, European Patent Office, www.epo.org.

new section "Y" of the ECLA. The primary purpose of this section was to tag documents related to new technological developments. Nanotechnology, which was the first emerging technology, was tagged in Section Y under subclass Y01N.[3] A new IPC/ECLA "B82Y" scheme superseded Y01N codes. The B82Y nanotechnology subclass is divided into nine main groups, eight of which relate to specific areas of nanotechnology. Since the beginning of 2011, patent searchers have been able to use the "B82Y"subclass to find documents relating to nanotechology in the world's patent databases. The 170,000 documents published before January 2011 and tagged using Y01N symbols were transferred to the corresponding symbols in B82Y.[4] Table 3.1 provides the details of the B82Y nanotechnology subclass.

The use of a combination of key words and the classification can be used effectively to truncate searches and enhance search effectiveness. For example, if "DNA computers" are used as keywords for the search, it will produce a large number of hits that may not be relevant to nanotechnology as there will be hits related to computing DNA sequence from sequencing data. To enhance effectiveness of a search by eliminating irrelevant hits, the unwanted (irrelevant) hits can be reduced by combining B82Y class in the ECLA search.

3.1.2 United States Patent and Trademark Office (USPTO)

The USPTO established an informal nanotechnology classification Class 977 (Digest I) in October 2004 and later expanded it to a cross-reference collection with over 250 new subclasses. Particularly for tagging nanotechnology patents, USPTO has chosen an approach also based on the notion of nano-structure and imposing a bottom limit of 1 mn. Class 977, Nanotechnology provides disclosures related to nanostructure and chemical compositions of nanostructure; devices that include at least one nanostructure; mathematical algorithms, e.g., computer software, specifically adapted for modeling configurations or properties of nanostructure; methods or apparatus for making, detecting, analyzing, or treating nanostructures; and specified particular uses of a nanostructure.[5] The technology centers (TC) in the Class 977 include TC 1600: biotechnology and organic chemistry; TC 1700: chemical and materials engineering; TC 2100: computer architecture software and information security; TC 2600: communications; TC 2800: semiconductor, electrical, optical systems; TC 3600: transportation, construction, electronic commerce; and TC 3700: mechanical engineering, manufacturing, and products.

Class 977 Digest I (Oct. 2004) provides a cross-reference art collection of 263 new subclasses. The creation of cross-reference Class 977 for nano-technology and its expanded 263 subclasses provides the USPTO with a consolidated area of search to supplement the patent application examination process as an enhanced search tool.[6] It is possible to combine text query and a cross-reference class 977 search. For example, the query could be "device" and "class 977, subclass 931." A query such as "device and ccl/977/931" may be entered in the advanced search option of USPTO patent search.

3.1.3 Japan Patent Office (JPO) Website

Since 1999, the JPO offers free search of its national patent collection via the Industrial Property Digital Library (IPDL) portal. IPDL has been operated by the National Center for Industrial Property Information and Training (INPIT) since October 2004.

A first possibility is to search Japanese patent publications by using the JPO internal classification, comprising the so-called "File Index" (FI) system and the "File Forming Terms" (F-terms) system. Other details useful for the search (e.g., database organization and syntax) can be found in the IPDL website (http://www19.ipdl.inpit.go.jp/PA1/cgi-bin/PA1INIT?1214892045792). The website is operated by INPIT (http://www.ipdl.inpit.go.jp/homepg_e.ipdl), an independent administrative institution of the JPO.[7]

English abstracts data named "PAJ" (Patent Abstracts of Japan) produced by the JPO is a database of English abstracts of unexamined Japanese patent applications that can be accessed in the IPDL website. Approximately 400,000 patents a year have been included in the database since 1976.

The dedicated classification systems for nanotechnology in IPC, EPO, USPTO, and JPO do not necessarily cover all patents related to nanotechnology. One may have to refer to other classes to identify the patents in fields that may not be searched by merely using the dedicated class. Some of the mature technologies of the past such as zeolites, which are crystalline materials with structured channels of 0.3 to 0.9 nm, may not be identified with the use of the dedicated nanotechnology classification system. It is well known that zeolites are synthesized using a method of self-assembly around template molecules. Their well-defined pore system with channels of molecular dimensions provides them with unique properties in catalysis and separation technologies. They are also referred to as "molecular sieves." This field, which had its beginning around the early 1950s and has matured to a major industry today, ought to have been considered as a field in nanotechnology and should have figured in the classification system.[3] The patent classification system is indispensable for the retrieval of patent documents in a search to establish the novelty of an invention or to determine the state of the art in a particular field of technology to establish inventive step.

The commonly used databases for patent searches are USPTO, EPO, JPO, WIPO, INPADOC, and Indian Patent Office. There are other free patent search websites such as www.freepatentsonline.com, www.google.com/patents, and www.patentstorm.com.

However, there are paid databases such as World Patent Index (Derwent World Patents Index from Thomson Scientific), MicroPatent, and STN as illustrated in Figure 3.1.[8]

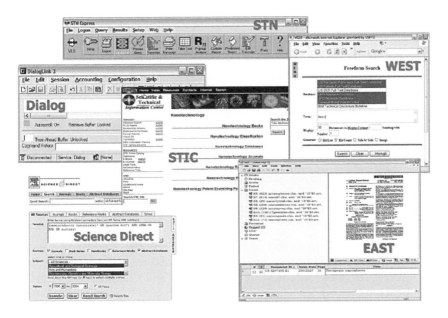

FIGURE 3.1
Search tools for nanotechnology.

3.2 Searching Other Sources for Nanotechnology Information

Along with patent search, it is important to conduct searches in other databases to identify relevant scientific publications.[9] The Web of Science, Scopus, CSA Illumina, and Google Scholar are additional general information sources. In their review, Huang, Notten, and Rasters provide a list of journals publishing nanotechnology articles and search strategies employed.[10]

Some other sources include

- Abstracts and indexes: These contain abstract and index (bibliographic) information of a given article. Some of the examples are MEDLINE (medicine), TOXLINE (toxicology), BIOSIS (biology), COMPENDEX (engineering and technology), and ERIC (education).

- Commercial full-text journal articles and digital libraries: Many commercial publishers have made their online content available on the Web. Prominent example is Web of Science, a product of Thomson Scientific. IEEE Computer Society provides online access to IEEE journals.

- Free full-text articles and e-print: This movement was initiated by the academic community to provide free access to journals and books. For example, *HighWire Press* is a service affiliated with Stanford University (highwire.stanford.edu/lists/freeart.dtl); Free Medical Journals site (www.freemedicaljournals.com); and the e-print service arXiv.org, supported by Cornell University, provides open access to e-prints in physics, mathematics, computer science, and quantitative biology.

- Citation indexing system and services: These provide aggregated and individualized citation information. Examples are the *Science Citation Index* (scientific.thomson.com/products/sci/), a product of Thomson Scientific; Google Scholar (scholar.google.com); and CiteSeer (citeseer.ist.psu.edu/citeseer).

- Electronic theses and dissertation (ETD): University Microfilms (UMI) has archived dissertations and master's theses with its ProQuest system (proquest.com). The Networked Digital Library of Theses and Dissertations (NTDLTD, www.ndltd.org) is another such resource.

- Business and industry articles and reports: Industry- and technology-specific reports are available at sites such as Forrester (www.forrestor.com), IDC (www.idc.com), and Gartner (www.gartner.com).

There are several websites, forums, blogs, social networking sites, open ware information centers, etc. that provide information on nanotechnology.

However, their quality and authenticity vary and therefore caution has to be exercised to cite them as credible prior art.

3.3 Creating Search Strategies

To arrive at relevant prior art, and to enhance the effectiveness of a search, one has to devise a structured approach and strategy in the context of the subject matter. Creating an appropriate search strategy involves the appreciation of the subject matter, the fields associated with the subject matter, definition of the problem, and the purpose for which the search is to be conducted. It is desirable to involve the inventors, personnel from marketing, and business strategy with the search specialist at the stage of conceptualizing and constructing the search strategies based on the search objective.

Suppose the invention relates to a gas sensor employing zinc oxide hexagonal crystals (or nanorods) deposited with metal nanoparticles. The problem addressed is to reduce the response time and enhance sensitivity operating at a reasonably lower temperature of the order of 250°C. If the search is conducted merely using the keyword "Gas Sensor" or "Sensor," it will result in retrieval of documents related to all kinds of sensors. However, inclusion of a keyword such as nanorods would focus the search at least to gas sensors with some linkage to nanorods. The search can be narrowed further with the inclusion of the keywords "metal oxide" and "crystal" depending on the purpose of the search.

For a quick search, the search strategy may be structured with the appropriate keywords in combination with the classification to achieve the highest probability of retrieving prior art.[11,12] The results of such a quick search helps in configuring further search strategies to achieve convergence in minimal steps of the iterative search process.

3.3.1 Types of Patent Searches

The search types are generally categorized based on the search objective as informative searches, freedom to operate or infringement searches, validity searches, and patentability searches.[13] The individual results of such searches may be used or may be appropriately combined to meet the set objectives.

The informative search is typically used for R&D planning, technology trend watch, country or enterprise statistics, forecasts of technological progress, competitors analysis, etc.[14,15]

The objective of carrying out freedom to operate or infringement searches is to determine whether valid patents in the subject matter of interest exist in a particular jurisdiction. This activity involves the identification of the relevant patents, their assignees, their legal status, and detailed analysis of the claims in relevant patents. The features of the products to be introduced into markets in specific jurisdictions are compared and contrasted against the claims of the relevant patents in those jurisdictions to access whether the products are infringing the relevant patents. Further, such an analysis provides insight into the technological core/strengths of the various patent applicants, which can be used effectively in charting business deals such as cross licensing, merger, acquisition, etc. Failure to locate patent claims infringed by the product or process may result in legal actions and high financial losses.

The purpose of a validity search is to assess the patentability of targeted patents. Based on the search, strategies are created to oppose them pregrant or postgrant or invalidate in revocation proceedings. In such a search, the relevant prior art is searched and analyzed with the perspective of establishing weaknesses in the "claimed" invention in terms of it lacking in novelty, inventive step, and industrial applicability (usefulness).

A patentability search is conducted to evaluate or establish novelty or inventiveness of an invention.

3.4 Illustrative Example

The example illustrates how a patent search is to be conducted for inventions related to metal nanoparticle antimicrobial paint preparations. Further, the problem is to understand the technical and patent landscape of the field of nanoparticles for antimicrobial effects and then develop a novel nanotechnology-based product and process for antimicrobial paints.

Step 1: Formulate the problem statement to appreciate the technical problem being addressed by the invention. Prepare a technology map based on the patents in the prior art and then carve out an approach that is non-infringing and further patentable.

List the essential technical features of the solution in the context of the invention.

Problem statement: The technical problems addressed in this invention are

- Provide compositions to avoid agglomeration of metal nanoparticles during mixing and also on aging of the paints in which they are incorporated.
- Avoid deterioration of the paint color with time.

Solution:

- Provide water miscible compositions containing silver nanoparticles and their tailored aggregates.
- Provide method of preparation of such aggregates that can be incorporated into paints during the preparation of the paints.

The invention exploits antimicrobial property of silver. The challenge is to produce the silver nanoparticles and their tailored aggregates that do not agglomerate in paints and meet the objects of the invention as given in the problem statement. This approach clearly avoids several paths that have been taken by various workers in the prior art. This has been arrived at from the technology map created analyzing results of extensive patent searches in the field of nanoparticles and their antimicrobial effects.

3.4.1 Search Strategies and Approaches

We illustrate here two search strategies and approaches, the first one carried out by the authors in consultation with Professor Joydeep Dutta of Asian Institute of Technology based in Thailand and the other conducted by the search experts from Sci-Edge Information, SciFinder/STN/Chemical Abstract Service (CAS) representative in India.

3.4.1.1 Strategy and Approach of the Authors

The keywords used are antimicrobial paints, resins, coats, emulsions, and antibacterial. The IPC identified are C09D5/16; C09D5/14; C09D5/14; C09D5/16; C09D175/04; C09D175/04; (IPC1-7): C09D5/16; C09D175/04; B32B5/16; B01J13/00C22B11/00, B82B.

Using this search concept, a technology map (illustrated in Figure 3.2) was prepared. From this technology map, it was decided to follow the route of "precursor dissolved in a solvent" as the most promising. This approach would be most compatible to the paint preparation process and would not involve any extra equipment or facility. Some process ingenuity would have to be devised for the preparation of the nanoparticles and their tailored aggregates and for their incorporation in paints. The search was then directed using silver nanoparticles, silver particles, silver colloid, and metallic nanoparticles as keywords.

The narrowed search provided the most relevant 35 patents and patent applications in the context of the problem as defined previously.

The following patents in groups were identified:

Group 1: Antimicrobial Coatings Resins
JP2006142292, US 2006009935, JP2008215618, US 20090136742, US 20080160264

Group 1a: Method of Preparation of Antimicrobial Resin
WO/2009/078618

Group 2: Antimicrobial Coatings Other Than Resins
US 20070224288, US 7476698, EP1427453, WO/2003/024494, JP3183766, JP07133445

The next step is to analyze the patents identified as most relevant prior art.

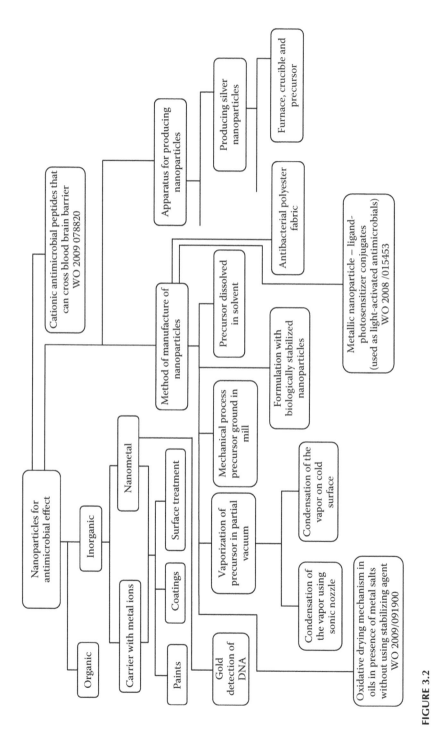

FIGURE 3.2
Technology map.

Group 1: Antimicrobial Coatings Resins
JP2006142292

- Coating method using a synthetic resin mixed with silver powder.
- Prevent nano-silver particles from crushing during process of mixing with resin resulting in precipitation of the nano-silver particles resulting in nonuniform distribution of the silver nanoparticles.

Technology used:

- Thermoplastic resin mixed with nano-silver powder by dispersion using ultrasonic distributor.
- Periodic pumping of the mixture to prevent precipitation.
- Coating applied on tiles.

US 20060099352

- Antibacterial and antistatic hard coating composition usable for a transparent plastic base material.

Technology used:

- Compounding of silver nanoparticle solution with a conductive filler, a photo-hardening resin, a photopolymerization initiator, and an organic solvent.

JP2008215618

- Suppression of detrimental bacteria (virus) in the water flowing through the cast iron pipe.

Technology used:

- Formation/deposition of a film of synthetic resin containing nano-silver particles on the inner side face of the cast iron pipe. Further, deposition of secondary coating of nano-silver particles on the resin film.

US 20090136742

- To impart superior antibacterial activity, corrosion resistance, conductivity, and adhesion to a steel plate.
- Nonconductivity of resin deteriorates weldability of steel; this problem is overcome by reducing thickness of film. However, reduction of film thickness results in a decrease in corrosion resistance.

Technology used:

- Antibacterial resin composition comprising aqueous silver containing solution.
- Aqueous silver containing solution comprising nano-sized silver particles 1 to 2 nm in diameter with concentration of 200 to 100,000 ppm; pH of the solution is 6 to 8.5; stabilizer in 0.5 to 1.5% by weight; anionic part of silver salt 1% by weight.
- Aqueous antibacterial resin composition comprising 100 parts by weight of at least one resin (selected from a group of acrylic, urethane, epoxy, and ester); 0.05 to 5 parts by weight of curing agent and aqueous silver containing solution so that silver concentration is 50 to 100 ppm.

US 20080160264

- Impart antifungal and antipest effect in decoration panel (for furniture) that comprises a pattern sheet and surface sheet impregnated with melamine resin.

Technology used:

- Impregnation of surface sheet with mixture solution consisting of melamine resin solution and a silver nano solution.
- Mixture consists of 0.2 to 0.5 wt% of the silver nano solution with respect to 99.5 to 99.8 wt% of the melamine resin solution.

Group 1a: Method of Preparation of Antimicrobial Resin
WO/2009/078618

- To impart antibiotic,* antistatic,† and electromagnetic wave shielding properties to resin.
- Preparation process becomes complex and uneconomical due to the requirement of dispersing agent and physical force (such as extruder, stirrer, melter, etc.) because polymer resin composition (obtained from a monomer) is dispersed in colloidal solution or solid powder of nano-sliver particles. Further, this imposes a limitation on the uniformity of nanoparticles dispersed in the resin composition.

---

* Applications of antibiotic resins: container for food and beverages, water-supplying pipe, refrigerator container, medical container, humidifier water tank.
† Applications of antistatic resin: preventing static electricity in home appliances.

Technology used:

- Precursor (organic silver complex that initiates formation of nano-silver particles) is dissolved in monomer (and not previously synthesized polymer resin) that is to be polymerized to resin composition, wherein organic silver complex is decomposed to form the nano-silver particles during radical polymerization[*] (contrary to the conventional process of dispersing nano-silver particles in previously synthesized polymer resin).

Group 2: Antimicrobial Coatings Other Than Resins
US 20070224288

- Obviate problems related to infection and bad odor due to bacteria or fungi in pedicure tub/spa.

Technology used:

- Application of antibacterial gel coating on tub surface.
- Antibacterial gel comprising industrially applicable gel mixed with silver nanoparticles 10 to 35 nm in diameter in a concentration of 50 to 200 ppm.

US 7476698, EP1427453, WO/2003/024494

- The application of antimicrobial adhesive and coating[†] for medical, technical-medical, and technical-hygienic applications is limited due to failure of the material in the inhibition zone measurement test (wherein material to be tested is kept in a culture medium such as agar) wherein due to antimicrobial acting metallic ions in an inhibition zone form around the material. The formation and size of this zone is the indicator of antimicrobial effectiveness. Conventional materials release an excessive concentration of silver ions.
- For medical applications, materials that show an inhibition zone are not suitable because such an excessive amount of silver is cytotoxic in nature and approval for the same is given by certifying authorities in accordance with the pharmaceutical laws, which is extremely expensive and time consuming.

---

[*] Radical polymerization method includes any of bulk, suspension, solution, or emulsion polymerization.
[†] Antimicrobial adhesive and coating materials are primarily synthetically manufactured material with an organic basis wherein the material hardens after processing.

- The challenge is to develop an antimicrobial adhesive and coating that does not indicate an inhibition zone in an inhibition test and hence is not cytotoxic.

Technology used:

- Antimicrobial adhesive and coating consisting of metallic silver particles in less than 5 ppm of silver, sodium, and potassium ions.
- Silver particles are made of aggregates of primary particles with an average grain size of 10 to 150 nm, preferably in the range of 80 to 140 nm, wherein silver particles (with aggregates with an average grain size of 1 to 20 µm, preferably 10 to 20 µm; surface area 3 to 6 m^2/g; porosity between 70 and 95%) created from such primary particles that are connected together are not cytotoxic.
- 0.01 to 5.0 wt% of silver particles are added in liquid organic component or additional organic component consisting of acrylate or methacrylate as an essential constituent, epoxide, urethane, silicone, or cyanacrylate

JP3183766

- To obtain a composition capable of forming a coating film having excellent antimicrobial properties, mildew-proof properties, and algicidal properties for a long period of time and with no change of color.

Technology used:

- The antimicrobial coating compound composition comprising 0.1 to 25 wt%, preferably 0.1 to 15 wt% of the fine particles based on the total of the fine particles and a film-forming agent.
- The antimicrobial metal component is formed into the fine particles in the form of a mixture or a compound with the inorganic oxide or the antimicrobial metal component is linked to the surface of fine particles of the inorganic oxide.
- Antimicrobial metal is selected from silver, copper, zinc, tin, or lead.

JP07133445

- To provide an antimicrobial coating having high antimicrobial activity, effective for preventing the color change and useful as a tray for holding an item such as perishable food.

Technology used:

- The composition comprising amorphous calcium phosphate particle supporting antimicrobial metallic ion such as gold and silver.

3.4.1.2 Strategy and Approach by Sci-Edge Information India SciFinder/ STN/Chemical Abstract Service (CAS) Representative in India

The prior art search was repeated for the same problem by information experts of Sci-Edge Information, who formulated the prior art search strategy/concepts and carried them out using the Science and Technology Network (STN) database. The authors acknowledge the inputs and help from Makarand Waikar and Dr. Jyoti Singh of Sci-Edge Information, SciFinder/ STN/Chemical Abstract Service (CAS) representative in India.

As indicated in the previous section, the search done by the nonexperts resulted in 35 patents/patent applications.

STN experts reformulated the prior art search strategy.

A common technique is to utilize the unique indexing present in the value-added databases with respect to substance and nanotechnology indexing terms as well as patent classification codes. The search expert first locates highly relevant records to evaluate their indexing and then expands the search with these new concepts. This methodology is also called a "pearl growing" strategy. The search area was segregated into four main concepts: silver, nanoparticles, antimicrobial, and paint. In order to be comprehensive, they explored the Internet to locate similar terms for each of the four concepts and online thesauri. These structured and indexed databases have respective online thesauri to help the users to know the evolution of a specific keyword over a period of time. For example, nanoparticles as a concept became a catchword from 1997 onward. Prior to 1997, such particles were known as "ultrafine particles."

```
=> FILE CAPLUS
=> E NANOPARTICLES/CT
E# FREQUENCY AT      TERM                    CT - Control Terms
-- ---------- --     ----
E1  0         2      NANOPARTICLE SUSPENSION/CT
E2  0         2      NANOPARTICLE SUSPENSIONS/CT
E3  160979    20 --> NANOPARTICLES/CT
E4  0         5      NANOPARTICLES (L) CORE-SHELL/CT
E5  0         4      NANOPARTICLES (L) FERROMAGNETIC/CT
E6  0         4      NANOPARTICLES (L) MAGNETIC NANOPARTICLES/
CT
E7  0         5      NANOPARTICLES (L) NANOCLUSTERS/CT
E8  0         5      NANOPARTICLES (L) NANODROPLETS/CT
E9  0         5      NANOPARTICLES (L) NANOPOWDERS/CT
E10 0         1      NANOPHASE/CT
E11 0         2      NANOPHASE COMPOSITE/CT
```

```
E12  0          9      NANOPHYES/CT              HNTE - History Notes
                                                 UF - Used for
=> E E3+ALL                                      NT - Narrower Terms
E13  81642      BT1    Particles/CT              RT - Related Terms
E14  160979     -->    Nanoparticles/CT
                       HNTE Valid heading during volume 126
                       (1997) to present.
E15             OLD    Particles (L) nano-/CT
E16             OLD    Particles (L) ultrafine/CT
E17             UF     Nanoparticle/CT
E18             UF     Nanoscale particle/CT
E19             UF     Nanoscale particles/CT
E20             UF     Nanosize particles/CT
E21             UF     Nanosized particles/CT
E22             UF     Ultrafine particles/CT
E23  14888      NT1    Pharmaceutical nanoparticles/CT
E24  896        NT2    Pharmaceutical nanocapsules/CT
E25  708        NT2    Pharmaceutical nanospheres/CT
E26  1950       RT     Mesophase/CT
E27  53219      RT     Nanocomposites/CT
E28  6817       RT     Nanocrystalline metals/CT
E29  32383      RT     Nanocrystals/CT
E30  1794       RT     Nanofluids/CT
E31  14961      RT     Nanostructured materials/CT
```

Additionally, a few other related terms such as nanocomposite and nano-crystals were identified.

In addition to the online thesauri, published literature is also classified under specific technical areas depending on the core research disclosed. In the case of patents, various technology areas are classified under IPC codes. In the case of journals, they are classified under classification codes.

A few classification codes related to a search in COMPENDEX database include

- 813.2: Coating Materials
- 761: Nanotechnology
- 547.1: Precious Metals

Some databases assign roles to substances discussed in the publication. For example, CAPLUS database assigns roles to all the compounds discussed in the paper/patent. Thus, if silver were discussed with respect to nanotechnology, its CAS Registry number would be assigned a role as NANO.

```
=> FILE CAPLUS
=> S 7440-22-4/NANO
L1    12578 7440-22-4/NANO
```

Therefore, one infers from the above illustration that there are over 12,500 references where silver metal has been discussed with respect to nanotechnology.

When all four concepts are worked on, it was possible to identify the associated terms for each concept.

Concept 1. Silver

> Silver, Ag, and any other terms that can be attributed to silver can be included in the search.

Concept 2. Nanoparticles

> Some of the similar key terms for nanoparticles are nanocluster, nanopowder, nanocrystal, ultrafine particle, and nanocomposite.

Concept 3. Antimicrobial

> Some of the similar key terms for antimicrobial are microbiocide, microbiocidal, microbiostatic, antibacterial, antifungal, fungicidal, fungicide, and antiprotozoal.

Concept 4. Paint

> The concept paint can be represented as paint, aerosol, emulsion, varnish, shellac, lacquer, resin, and coat.

Because this search would essentially be a keyword search, one needs to exercise caution and consider all possible variants of the search terms; for example, ultrafine particles may be expressed as "ultra fine particles" or as "ultrafineparticles."

3.4.1.2.1 *Arriving at the Final Search Strategy*

Step 1—INDEX feature was used to identify which databases would give the best hits. Once the relevant databases were identified, each was studied in detail to determine their coverage, content, etc. to decide if the database needed to be searched. If found irrelevant, the database was dropped from the search. For example, the database Aerospace deals with data pertaining to aerospace technology and was considered to be irrelevant in the present search.

```
=>INDEX CHEMISTRY
=> S NANOPARTICLE AND ANTIMICROBIAL

=> D RANK
F1    2198    CAPLUS
F2    1145    SCISEARCH
F3     590    COMPENDEX
F4     300    PASCAL
F5     267    INSPEC
F6     187    RAPRA
F7     137    WSCA
```

```
F8     100    CABA
F9      73    METADEX
F10     46    AGRICOLA
F11     36    DISSABS
F12     32    CBNB
F13     27    CEABA-VTB
F14     27    IPA
F15     16    BABS
F16     14    KOSMET
F17     13    CERAB
```

Step 2—Once, the databases were selected, each of the relevant databases was searched for all four concepts to arrive at separate answer sets.

Step 3—The four answer sets were critically studied by appropriately sampling a few records to get additional insights on how these concepts are indexed.

Step 4—The additional concepts were added to the search strategy after reviewing a few records obtained in Step 3. When several databases are searched together, one ought to be able to recognize that each database has its specific database feature, which must be taken into consideration for comprehensive and effective search strategies. For example, INSPEC and COMPENDEX have their own way of classifying concepts as classification codes (CC). Similarly, CAPLUS has specific roles assigned to the compounds. Hence, NANO is a role that is appended to the compound which is in a nano form or as nanoscale. One may explore them and check if the concepts have any specific numeric codes that can be used. While doing so, one may need to handle these databases separately.

Step 5—Concept 1 and concept 2 are combined with (5A), indicating that these two concepts can be apart from each other to a limit of 0 to 5 words between them. Here, (5A) is "Adjacent Proximity Operator," indicating that two concepts can be adjacent, in any order, and with 0–5 words between them.

Step 6—Concept 3 and concept 4 are combined with (5A), indicating that these two concepts can be apart from each other to a limit of 0 to 5 words between them. Here, (5A) is "Adjacent Proximity Operator," indicating that two concepts can be adjacent, in any order, and with 0–5 words between them.

Step 7—The results of Step 5 and Step 6 are combined using "AND."

Step 8—The search results are combined and the duplicates are removed.

Step 9—Finally, the answers retrieved can be displayed to study and analyze them with respect to the invention, patent assignees, year of publication, and claimed invention.

IPC codes may be included in the search. Although nanotechnology-related inventions are categorized in a specific IPC, it is not necessary that all the nanotechnology-related inventions are categorized only in this classification. Due to the ubiquitous nature of nanotechnology applications, related inventions may be classified in other classifications.

When the search was done following the above strategy and inputs, 117 answers were obtained that included patents and nonpatent literature,

English and non-English literature, etc. Some of the databases that gave the relevant hits in the context of the problem searched were CAPLUS, COMPENDEX, RAPRA, and WSCA.

The benefits of using structured databases in STN include value-added and structured data, one stop source and single search of multiple databases, non-English literature, search by structure/reaction/sequence codes, and reporting and patent mapping tools. Additionally, one can use features such as AnalysePlus and AnaVist to get unique insights into trends and patterns of the research areas of interest.

These search results were categorized based on various aspects as follows.

Table 3.2 indicates the occurrence of the results with respect to the language wherein 75% of the documents are in English, 9% are in Chinese, and so on. Very few documents in languages other than those mentioned in Table 3.2 account for a fraction of a percentage. These are not included in the table.

Table 3.3 indicates the number of documents out of searched documents published from the organization/assignee with respect to the keywords (nanoparticles, antibacterial agents, etc.) mentioned in the column. The list provided in Table 3.3 is only a representative list for getting a fair idea of the publication trend from organizations. It is not an exhaustive list comprising all the segregated documents.

Table 3.4 indicates out of the searched documents, patents from respective countries on a percentage-wise basis.

Table 3.5 provides the patent assignees based on the searched documents.

Table 3.6 provides the publication year trends based on the searched documents. Although not exhaustive (comprising 117 documents), it provides a fair idea of the publication trend.

Analysis with respect to countries and the publication year is depicted in Figure 3.3. The research landscape illustrated in Figure 3.4 represents the peaks corresponding to the number of times a particular word (such as *varnish*) is discussed in the searched documents. Height of the peak is proportional to the occurrence of the particular term in the searched document.

TABLE 3.2

Analysis on Language of the Original Document

	Occurrence (%)	Language
1	75	English
2	9	Chinese
3	5	Italian
4	5	Russian
5	2	German
6	2	Japanese
7	2	Persian

TABLE 3.3

Key Organizations by Technology Indicators

Key Organizations/ Assignees by Technology Indicators	Nanoparticles	Antibacterial Agents	Coating Materials	Paints	Antimicrobial Agents
Tsinghua University	3	3	1	2	0
AcryMed, Inc.	2	0	1	0	2
Thermolon Korea Co. Ltd.	2	2	2	2	0
Instytut Chemii Przemyslowej im. Prof. I. Moscickiego	1	0	1	2	2
Northwest Normal University	0	0	2	1	0
Central South University	2	1	1	1	0
Hong Kong Polytechnic University	2	2	0	0	0
Procter & Gamble	1	1	0	0	1
Advanced Medical Solutions Limited	0	0	1	0	1
Centro de Investigacion en Quimica Aplicada	1	1	1	1	1
Yamagata University	1	1	0	0	1
Ohara Kinzoku Kogyo K. K.	1	0	0	0	0
Gdansk University of Technology	1	1	0	1	0
University of Tehran	0	1	0	0	0
Nanoco SP Z O O	1	0	0	0	1
Freudenberg & Co.	1	0	1	0	0
Coolleader Co. Ltd.	0	0	1	0	0
Loggia Industria Vernici Srl	1	1	0	1	0
Hanoi University of Technology	1	1	1	0	0

3.4.1.2.2 Use of These Search Results

Professor Joydeep Dutta and colleagues of the Asian Institute of Technology based in Thailand planned their work to clearly avoid the prior art process and developed an innovative process in solution wherein the silver particles were precipitated using salts of weak organic acids in the presence of poly acids and surfactants further adding a base followed by reduction. In particular, the researchers developed controlled precipitation of silver salts using trisodium citrate in the presence of polyacrylic acid and sodium

TABLE 3.4

Analysis Results by Patent Country

	Patents (%)	Patent Countries
1	32	CN
2	27	KR
3	19	US
4	16	WO
5	14	EP
6	8	JP
7	7	AU
8	7	DE
9	4	RU
10	3	AT
11	3	CA
12	3	MX
13	1	ES
14	1	PT
15	1	BR
16	1	IN
17	1	IT
18	1	JP
19	1	NZ
20	1	PL
21	1	TR

TABLE 3.5

Top 10 Patent Assignees

1	Tsinghua University
2	Northwest Normal University
3	Hong Kong Polytechnic University
4	AS Russian Academy of Science Institute of Theoretical and Applied Mechanics
5	Thermolon Korea Co Ltd
6	ItN Nanovation GmbH
7	Instytut Chemii Przemyslowej im. Prof. I. Moscickiego
8	Chinese Academy of Sciences
9	Central South University
10	AcryMed, Inc.

TABLE 3.6

Publication Year Trends

	Publication Year	Document Count
1	2011	12
2	2010	11
3	2009	4
4	2008	6
5	2007	9
6	2006	17
7	2005	9
8	2004	5
9	2003	5
10	2002	1
11	2001	1
12	2000	1

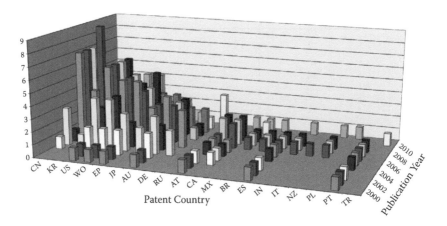

FIGURE 3.3
Analysis with respect to country and publication year.

dodecyl sulfate, adding NH_4OH and further reducing the same using $NaBH_4$ to provide an aggregate of silver nanoparticle with a size of at least 200 nm, wherein silver particle size is in the range of 4 to 30 nm, and filed a patent application.

The exercise presented above illustrates the potential of a well-structured literature search by a trained information scientist using comprehensive databases.

Other popularly used paid databases are World Patent Index (Derwent World Patents Index from Thomson Scientific) and MicroPatent.

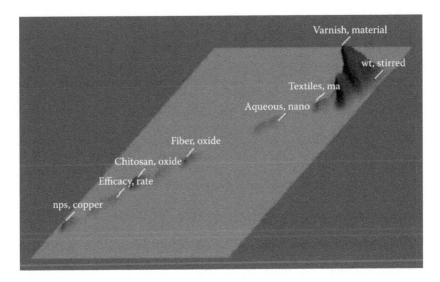

FIGURE 3.4
Research landscape.

References

1. IPC version 2009.01 available at www.wipo.int.
2. European Convention on the International Classification of Patents for Invention of December 19, 1954, which came into force and was published by the Secretary General of the Council of Europe on September 1, 1968.
3. Scheu, M., Veefkind, V., Verbandt, Y., Molina Galan, E., Absalom, R., and Forster, W., Mapping nanotechnology patents: The EPO approach, *World Patent Information*, Elsevier, the Netherlands, 2006, pp. 204–211.
4. Nanotechnology and Patents, European Patent Office, www.epo.org.
5. Nogueira de Sousa Branquinho Nordberg, A.R., Nanotechnology patents in Europe: patentability exclusions and exceptions, master's thesis, Stockholm University, 2009, http://www.juridicum.su.se/juruppsatser/2010/ht_2010_Ana_Rita_Nogueira_de_Sousa_Branquinho_Nordberg.pdf.
6. Kisliuk, B., USPTO Assistant Deputy Commissioner for Patent Operations, Nanotechnology Partnership Forum, NIST, Gaithersburg, MD, presentation received through personal communication, 2010.
7. Oda, S., JP-NETe—An English-language search tool for Japanese unexamined patents, *World Patent Information* 31, 131–134, 2009.
8. Nguyen, D.T., USPTO Supervisory Patent Examiner, Art Unit 1633, Biotechnology, Pharmaceuticals, Organic Chemistry, Presentation entitled "Update on nanotech-related initiatives and examination," www.cabic.com/bcp/061306/DNguyen_NU.ppt.

9. Chen, H. and Roco, M.C., *Mapping Nanotechnology Innovations and Knowledge, Global and Longitudinal Patent and Literature Analysis*, Springer Science + Business Media, LLC, New York, 2009.
10. Huang, C., Notten, A., and Rasters, N., Nanotechnology publications and patents: A review of social science studies and search strategies, United Nations University UNU-MERIT, Maastricht Economic and Social Research and Training Centre on Innovation and Technology, The Netherlands, Working Paper Series, 2008.
11. Michel, J. and Bettels, B., Patent citation analysis—a closer look at the basic input data from patent search reports, *Scientometrics*, 51(1), 185–201, 2001.
12. Prior art searching. Canadian Bacterial Diseases Network, http://www.cbdn.ca/english/ip_primer/Web/PriorArt-Searching.html.
13. Pasquale, F., Patentability search strategies and the reformed IPC: A patent office perspective, *World Patent Information* 29, 33–53, 2007.
14. Wu, C. and Liu, Y., Use of the IPC and various retrieval systems to research patent activities of U.S. organizations in the People's Republic of China, *World Patent Information* 26(3), 225–233, 2004.
15. Chakravarti, A.K. et al., Patent Information and electronics, *Electronics Information & Planning* 22(11), 559–570, 1995.

4

Patent-Led Nanotechnology Business: Perspectives

Nanotechnology has progressed from its conception to birth at a reasonable pace and during its crawling phase experienced basic teething issues like all pioneering technologies. The early signs of nano's attempts to stand on its own feet were seen in 1999 with the emergence of promising options involving first-generation nanomaterials, molecular manufacturing processes, tools, devices, and systems for a wide range of applications. Universities together with centers of higher learning and excellence immensely contributed to the growth of knowledge firing the imagination of several technology entrepreneurs, who began to visualize and intelligently anticipate opportunities beyond the horizon. It has been a fertile area for the seeding and germination of entrepreneurs setting up new business ventures. The universities in which scientists and technologists were working then offered to rent the rights to their patents on liberal terms to the techno-entrepreneurs to enable the developed knowledge to transit from the academic domain into a fuzzy commercial universe.

Nanotechnology start-ups sprouted in clusters and the commercial world began to sense business reality in a not-too-distant future. From mid-1990s, the field of nanotechnology has seen a variety of high technology patent-based startups ranging from university spin-offs, to government lab spin-offs, corporate spin-offs, to private entrepreneurs seeking technologies from diverse sources at least until they are able to generate their own sales and profits.

For example, Northwestern University Professor Chad Mirkin and his associates formed two such companies—NanoInk and Nanosphere—based on their research. NanoInk, Inc. received the first dip ink pen nanolithography patent, and Nanosphere, Inc., a life sciences nanotechnology company, holds several nanoparticle probe technology patents. On the corporate side, Nonvolatile Electronics, Inc. (presently called NVE Corporation) was a Honeywell spin-off nanotechnology company.

There have been diverse sources of funding ranging from raising capital from family and friends, and government funding, bank loans, angel investors, venture capital providers, and in some cases investments from large firms.

Governments around the world also stepped up funding in nano R&D and large industries initiated scouting for potential products and processes for adaption and integration into their existing operations and into new avenues for early commercialization. Joining the initial marchers like governments in the United States, Germany, France, United Kingdom, and Japan, were new kids on the block such as China, South Korea, Taiwan, and India leading to an irreversible transformation in the demography of nanotechnology from early 2000.

In emerging fields, such as nanotechnology, in addition to patent rights, businesses exploit trade secrets especially in areas in which reverse engineering of the technology is not easy. The patents are drafted to make adequate enabling disclosures. However, some of the technology knowledge is held back as trade secrets/know-how. The two are traded with the appropriate patents and trade secrets in an interwoven manner to derive maximum value.

However, trade secrets are to be protected appropriately and a complete due diligence is to be conducted to ensure that the necessary conditions of trade secrets are followed for the information and knowledge to qualify as trade secrets.

The nano-reality space began to get densely dotted with patent applications indicated by a dramatic rise in the number of patent filings with a rapid shift in the ratio (R) of corporate nanotech patent applications to corporate nanotech publications in a short time. "R" of ~0.4 in 1999 grew to ~0.8 to 1.1 in 2003 and further to 1 to 1.5 in 2006–2007, indicating a surge in promising utilization of nano-concepts in novel products and processes.[1]

Another notable observation is the unusually large stake that universities have in nanotechnology. It is estimated that approximately 20% of nanotechnology patents are owned by universities. As patents resulting from upstream research generally have the potential to claim broad patents covering core building blocks needed to implement downstream nanotechnology applications, they have significant ramifications on the development and commercialization of nanotechnology-enabled products, devices, systems, and manufacturing processes.

As convergent applications based on nanotechnologies evolve, incremental innovations lead to the filing and granting of multiple overlapping patents from competing groups. Such a phenomenon results in regions of the high-density growth of protected domains in the nanotechnology patent landscape (nanoscape) known as "patent thickets." In the case of nanotechnology, the situation becomes complex due to the hybrid and ubiquitous nature of this technology wherein the granted patents have broad claims covering the results of several technology intersections, making the patent landscape murky with blurred delineation boundaries.

Platform technologies primarily developed for problems in one field can be applied to solve problems in other fields. For example, emulsions prepared with applications in the cosmetics industry may find use in the agrochemical industry for insecticide or fertilizer sprays or as drug delivery systems in nanomedicine. The patent claims for such inventions generally end up with broad claims on formulations for emulsions and become strategic tollgates creating barriers for follow-on innovators thereby impacting commercialization of a range of emulsions with applications in a spectrum of industries. This inclusive and cumulative nature of nano-inventions and patents has been the center point of techno-legal and ethical debates.

Commercialization in the field of nanotechnology is fraught with issues involving several patent minefields. Strategic planning therefore is essential before embarking on any business mission.

Some of the IP transaction models involving licensing and assignments by way of illustration before a patent holder to exploit his monopoly are

- Use the owned patent (after obtaining the necessary statutory approvals) to manufacture the product, market the product, or contract the marketing of the product to other parties.

- License (rent the right) the patent to third parties with or without specific field of use, to manufacture the product but market the product by self or by contracting marketing of product to same or other parties in any territory (or in specified territories) with mutually negotiated benefit-sharing arrangements.

- License (rent the right) the patent to other parties with mutually negotiated benefit-sharing arrangements to manufacture the product and leave it to the licensee to decide on the business model for marketing the product.

- Negotiate in-license or out-license with patented technologies for further R&D, manufacture, or marketing of product with specified geographical areas for specified activities.

- Barter rights by way of cross licenses for independent or collaborative working.

- Establish franchises involving other parties.

- Sell off the patent/assign the patent for an appropriate return.

- Set up a joint venture and merge patent portfolios under specified terms of their use.

- Enforce patent against infringers of the patent portfolio.

- Selectively let patent rights lapse in specific jurisdictions based on business and other considerations.

- Donate patents to institutions/individuals with or without conditions attached.

Nanotechnology is presenting a situation in which the densely packed patent thickets make it extremely difficult for the stakeholders to commercialize their patented technologies without trespassing into the protected domain of others. This creates a situation of multidependencies and therefore without each other's consent the developed technologies cannot be taken to the market without infringing upon each other's patents. Such a situation may be overcome through complex multiple-stacking of patent licenses/cross-licenses or through the creation of innovative "patent pools," which are contractual undertakings involving agreements between two or more patent owners to license one or more of their patents to one another or third parties. These aspects are well illustrated in a recent article by Sylvester and Bowman.[2]

The concept of "patent thickets" was beautifully illustrated by Clarkson and DeKorte in 2006[3] with their mapping of the nanotechnology patent landscape based on their study of the patents in the United States in 2000, 2002, and 2004. It is illustrated in Figure 4.1.

Around 2004, companies in the United States postured themselves according to their niche technical expertise and financial strength. The nanotech commercialization process was initiated in the United States in early 2000 as

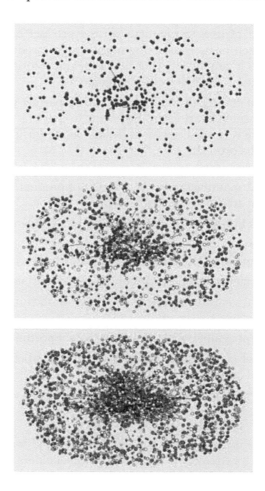

FIGURE 4.1
Growth of nanotechnology patents in 2000 (top), 2002 (middle), and 2004 (bottom). (From Clarkson, G. and Dekorte, D., *Ann. N.Y. Acad. Sci.*, 1093, 180–200, 2006. http://deepblue.lib. umich.edu/bitstream/2027.42/72678/1/annals.1382.014.pdf.)

shown by the set of operating companies with their area of business in parentheses, namely: NanoInk (lithography/production); Npoint and Veeco (manipulation/characterization); General Nanotech (CAD tools); Nanophase (bulk nanomaterials); Inframat (industrial coatings); Catalytic Solutions (catalysts); Nanotex (textiles); Nanosphere, C sixty, and Nanobio (life sciences); Molecular Nanosystems, Nanosys, Nantero, and Konarka (integrated systems). A set of large industrial houses also attempted to adapt, innovate, and integrate nanotechnology into their businesses such as Dupont (materials); IBM, HP, and Intel (computing); Roche (life sciences); Exxon Mobil (energy); Lucent (communications equipment); Hitachi (consumer electronics); and Raytheon (aerospace).[4] Some of these companies continue to be active and live in the nanobusiness landscape while expanding their patent portfolios in their respective areas of business.

4.1 Integration of a Fragmented Patent Landscape

After several decades of development, nanotechnology is in the process of maturation and consolidation. In this phase, the field will experience shake out of companies, mergers, and acquisitions to enable the creation of companies with the critical mass necessary for profitable sustenance in an aggressive competitive market. In many start-up companies, there has been an over-emphasis on creation of patent portfolios with the fond hope that anything new with nanotechnology would find an obvious market for commercialization. However, this has proved to be elusive as several start-ups produced too few results in terms of revenues. Nevertheless, valuation of the strategic patent portfolios and know-how is of immense significance to the companies engaged in this survival contest to not only establish and maintain technology dominant positions, but also to trade the intellectual assets even during restructuring processes in insolvency proceedings.

Nanotechnology has also seen several horizontal and vertical mergers/integration where the companies involved in similar kinds of product lines merge with each other. The principal objectives behind these types of mergers is to achieve the following: synergies in technological expertise, complementary patent portfolios, economies of scope and scale, avoid duplication of installations, fragmented services and functions, widen the line of products, decrease working capital and fixed assets investment, planned risk diversification, minimize competition, advertising expenses, enhance the market capability, ensure steady and cost-effective supply chain to service the emerging markets, and to get more dominance on the market without being accused of becoming monopolistic under competition laws or antitrust laws.

A few examples of nanotechnology start-ups illustrate approaches that have been successful in the past decade.

4.2 Case Study 1: NanoInk, Inc.

NanoInk, headquartered in the Illinois Science and Technology Park, north of Chicago, is a technology-led company specializing in nanometer-scale manufacturing and applications development for the life sciences, engineering, pharmaceutical, and education industries. Starting off with the founder technologist Professor Chad Mirkin and using Dip Pen Nanolithography® (DPN®), a patented and proprietary nanofabrication technology that involves deposition of nanoscale materials onto a substrate for applications in the life sciences and semiconductor industries, the company has developed a range of technologies to rapidly, easily, and cost effectively create micro- and nanoscale structures from a variety of materials on a range of substrates.

NanoInk's approach has been to develop applications internally in fundamental nanolithography, bioarrays, advanced materials, repair, and instruments with simultaneous in-licensing of patents from universities and other companies.

NanoInk has created a patent portfolio of over 250 patents and applications filed worldwide and licensing agreements with Northwestern University, Stanford University, University of Strathclyde, University of Liverpool, California Institute of Technology, and the University of Illinois at Urbana-Champaign. In 2006, NanoInk, Inc. and SII NanoTechnology Inc. (SIINT), a subsidiary of Seiko Instruments Inc., signed an exclusive licensing agreement to provide nanoscale repair solutions to the photomask industry. The two companies agreed to collaborate on projects that would involve modification of NanoInk's proprietary DPN technology to be integrated with SIINT's photomask repair instruments and nanomachining platforms. In 2007, NanoInk exclusively licensed a patent from Arrowhead Research Corporation, which was developed through Arrowhead's sponsored research at the laboratory of Dr. Patrick Collier at the California Institute of Technology. The idea was to strategically expand NanoInk's patent portfolio and to support applications of their DPN technology to drive the development of high value end products.

4.3 Case Study 2: Nanosphere, Inc.

Nanosphere was founded in 2000 based upon nanotechnology research by Dr. Robert Letsinger and Dr. Chad Mirkin at Northwestern University in Evanston, Illinois. They were able to achieve the functionalization of gold nanoparticles with oligonucleotides (DNA or RNA), or antibodies, for diagnostic applications to detect nucleic acid or protein targets, respectively.

To date, Nanosphere has been developing, manufacturing, and marketing an advanced molecular diagnostic platform, the Verigene System, which enables multiplexing, simple, cost-effective, and highly sensitive nucleic acid (DNA and RNA) and protein testing on a single platform to improve patient care at lower costs by providing time-critical information, allowing earlier detection of disease and more targeted treatment.

Nanosphere's patent portfolio is comprised of 151 issued patents and 52 pending patent applications, which either they own directly or for which they are the exclusive licensee. Some of these patents and patent applications derive from a common parent patent application or are foreign counterpart patent applications and relate to similar or identical technological claims. The issued patents cover approximately 11 different technological claims and the pending patent applications cover approximately four additional technological claims. Many of Nanosphere's issued and pending patents were exclusively licensed from the International Institute for Nanotechnology at Northwestern in May 2000, and they generally cover Nanosphere's core technology, including nanotechnology-based biodiagnostics and biobarcode technology. In addition, as of December 31, 2010, they had nonexclusive licenses for at least 47 U.S. patents that covered 12 different technological claims from various third parties.

Nanosphere's aggressive pursuit in these niche areas is very well supported by its close proximity and active link to the academic world.

4.4 Case Study 3: NVE Corporation

NVE Corporation (founded in 1989 as Nonvolatile Electronics, Inc. in Edina, Minnesota) is Honeywell's spin-off nanotechnology company. It is a very strong technology–led company as is evident from its patent portfolio of more than 50 U.S. patents issued and assigned to NVE by 2011. The company also has a number of foreign granted patents, a number of U.S. and foreign patents pending, and patents licensed from others. More than 100 patents worldwide issued, pending, or licensed from others protect NVE technology. All these support the business of pioneering devices such as sensors and couplers using spintronics to acquire, store, and transmit information, which are used in industrial, scientific, and medical applications. The company has licensed its spintronic magnetoresistive random access memory technology, commonly known as MRAM. NVE's products are sold through a worldwide distribution network. The authors have depicted the business model in Figure 4.2.

NVE's science and technology base continually is rejuvenated through its close collaborative links with the academic world as is evident from its

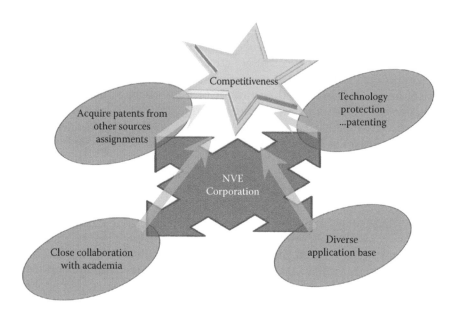

FIGURE 4.2
NVE business model.

recent joint publications with several universities in the United States and elsewhere. Some of their recent publications include

Ya. B. Bazaliy and A. Stankiewicz, Ballistic precessional contribution to the conventional magnetic switching, *Applied Physics Letters*, 98, April 4, 2011.

J. E. Davies et al., Reversal of patterned Co/Pd multilayers with graded magnetic structure, Conference on Magnetism and Magnetic Materials, November 15, 2010, Atlanta, GA, *Journal of Applied Physics*, 109, March 28, 2011.

J. Davies, A. Stankiewicz, and J. G. Deak, Asymmetric hysteresis loop expansion in strained magnetic tunnel junction, Conference on Magnetism and Magnetic Materials, November 16, 2010, Atlanta, GA.

D. A. Baker and J. L. Brown, Advances in solid-state magnetic sensors for medical devices, Advanced Technology Workshop on Microelectronic Packaging for Medical Devices, June 10, 2010, Minneapolis, MN.

A. Stankiewicz, Energetic analysis of fast magnetic switching, APS March Meeting, March 16, 2010, Portland, OR, *Bulletin of the American Physical Society*, 55(2).

J. E. Davies, B. J. Kirby, K. Liu, S. M. Watson, G. T. Zimanyi, R. D. Shul, P. A. Kienzel, and J. A. Borchers, Controlled anisotropy variation in cobalt thickness graded cobalt/palladium multilayers, Joint MMM-Intermag Conference, January 22, 2010, Washington, D.C.

This is a very typical model of a high-tech company in which the company simultaneously carries on its R&D activities to develop diverse applications based on its S&T base and aggressively protects them through patent applications to expand its patent portfolio. In addition to these activities, the company acquires patents from other sources through assignments.

4.5 Case Study 4: Vista Therapeutics, Inc.

Vista, headquartered in Santa Fe, New Mexico, was co-founded in 2007 by Spencer Farr, an experienced scientist and entrepreneur, and Charles Lieber, Distinguished Professor of Chemistry at Harvard University.

The technology involving nanowires permits continuous and real-time monitoring of multiple biomarkers in blood and urine using biosensors. Further functionalized nanowires can monitor and detect on a continuous and dynamic basis antibody-antigen interactions, enzyme substrate interactions, and gene expression.

Vista has created a model in which its team of world-class experts in medicine, biotechnology, nanotechnology, engineering, chemistry, and informatics develop cutting-edge technologies in their laboratory, continually collaborate in essential fields with Harvard University, and actively link up with pharmaceutical companies, doctors, and medical researchers for the development and deployment of its NanoBioSensors and is now approaching markets that do and do not require FDA approval. The technologies are aggressively protected by building and managing a strategic patent portfolio.

Vista signed license agreements with both Harvard University and Nanosys covering several patents and patent applications related to the use of nanowires for biosensors. Under the terms of the agreements, Vista secured the exclusive, worldwide rights for the use of nanowires for detection of biomarkers associated with organ or tissue damage, and any form of treatment or therapeutics-associated adverse responses. In consideration, Harvard and Nanosys received an equity position in Vista, as well as upfront license and downstream royalty payments. This allows Vista to commercialize through manufacture and sale of nanowires that are formatted to provide real-time, continuous measurement of blood and urinary biomarkers of organ and tissue injury.

It is instructive to know that in the year 2000, Professor Lieber made a patent application which resulted in the patent in the name of C. M. Lieber, T. Rueckes, E. Joselevich, and K. Kim, "Nanoscopic Wire-Based Devices and Arrays," U.S. Patent 6,781,166. After filing the patent in 2000, Professor Lieber

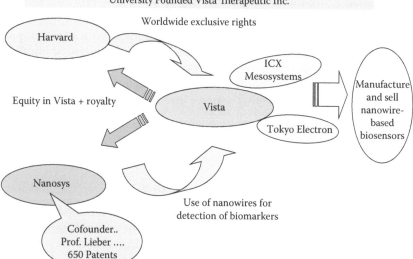

FIGURE 4.3
Vista business model representation.

and his colleagues published their seminal paper titled, "Single-Nanowire Electrically Driven Lasers."[5]

By filing the patent first and then publishing the findings in *Nature*, the inventors ensured that the novelty was not lost and that their own publications did not in any way jeopardize the patentability of their invention.

In June 2010, ActiveCare, Inc. and Vista created a strategic relationship to integrate Vista's nanowire technology in ActiveCare's products and services. The terms of the strategic agreement between Vista and ActiveCare are to make an equity investment in Vista Therapeutics and to pay a fee for the development of products utilizing Vista's nano biosensor technology. As part of the strategic agreement, ActiveCare will have exclusive use of Vista Therapeutic's technology in the elderly market (excluding hospitals, for which ActiveCare will have the right to acquire an exclusive sublicense).

The authors have created a visual representation of VISTA's business model. It is illustrated in Figure 4.3.

4.6 Case Study 5: Applied Nanotech Holdings, Inc.

The company was founded in 1987 and incorporated in Texas in 1989 as SI Diamond Technology, Inc. for field emission displays. It went public in 1993.

In 1996, it created Nano-Proprietary, Inc., a holding company consisting of two wholly owned operating subsidiaries: Applied Nanotech, Inc. (ANI) for CNT electron emission and Electronics Billboard Technology, Inc. for electronic billboards for digital advertising. In 1999, ANI licensed its FED technology. In 2005, it focused on carbon nanotubes technology, and in 2006, it diversified into other areas of nanotechnology and sold off Electronic Billboard Technology. From 2006 to the present, ANI organized its business in five divisions, namely, nanomaterials, nanoelectronics, nanosensors, nanoecology, and CNT electron emission.

ANI follows a three-pronged business model involving R&D services, IP licensing, and subsidiary or joint venture relationships. Such a model is possible to sustain only because of a strong focused patents portfolio in the business areas of the five divisions. Interestingly, ANI holds 126 issued patents in the United States and around 120 patents pending in the United States as is evident from the profile illustrated in Figure 4.4a–d. Besides their U.S. portfolio, ANI has filed and has patents granted to it in various countries around the world.

The patent estate of ANI has been segregated by the authors into nanomaterials (CNT composites, thermal management, and nanoparticles); nanoelectronics (nanoparticle inks, nanoparticle pastes, and Exclucent™ materials); nanosensors (metal nanoparticle sensors, enzyme-coated carbon nanotube sensors, sono photonic sensors, and ion mobility sensors); nanoecology (PhotoScrub®); and CNT electron emissions (electron emission activities, CNT display applications). Figure 4.4a–d clearly indicates the density of the practically impenetrable ownership of inventions in their competitive business areas.

ANI has recorded expenditure of $4,839,556 (64% of operating cost); $3,662,323 (59% of operating cost); and $4,614,644 (54% of operating cost) on R&D in the years ended December 31, 2010, 2009, and 2008, respectively. In 2011, the company was expected to spend over 60% of its operating cost.

This continual creation of directed intellectual property by ANI enables it to license the patents selectively to a partner or partners for product commercialization and to establish a recurring royalty stream. Its royalty agreement generally consists of both an upfront payment and an ongoing royalty based on sales of products using the technology. If a royalty agreement includes exclusivity for a particular technology or market, it also includes required minimum royalty payments. Further, multiple revenue streams are created through additional royalty generating license agreements for products shipped by licensees. ANI has no manufacturing facilities and does not intend to establish any such facilities on its own. However, ANI may participate in manufacturing operations with its partners through the use of subsidiaries, joint ventures, or other arrangements. ANI is a typical case of business through efficient and strategic management of intellectual property rights. In 2008 and 2009, it executed licensing agreements for CNT-enhanced epoxy and for nano-copper inks and pastes, respectively.

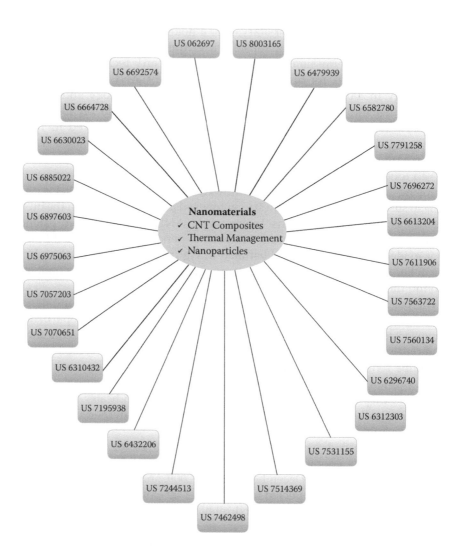

FIGURE 4.4a
ANI patent portfolio.

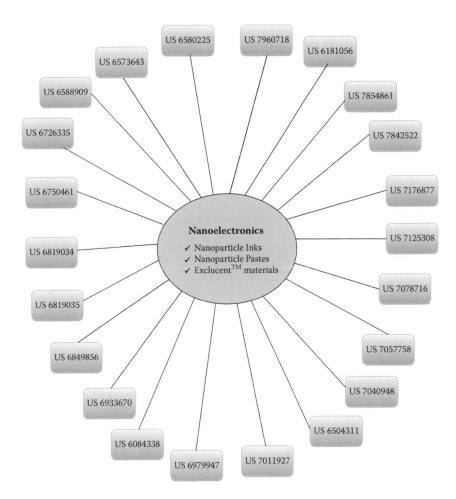

FIGURE 4.4b (*Continued*)
ANI patent portfolio.

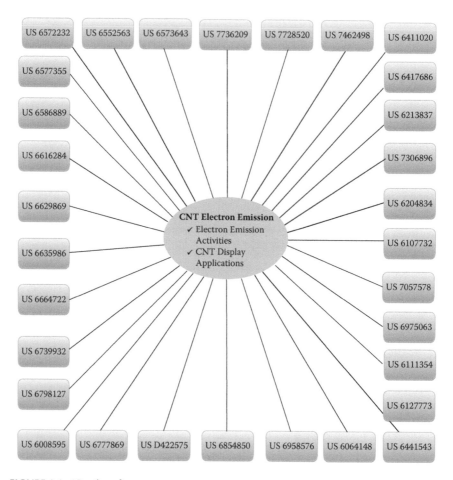

FIGURE 4.4c (*Continued*)
ANI patent portfolio.

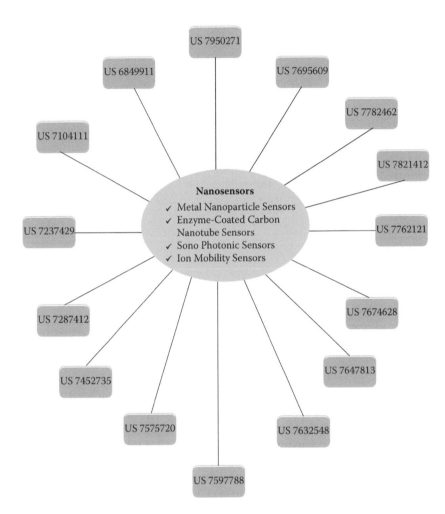

FIGURE 4.4d (*Continued*)
ANI patent portfolio.

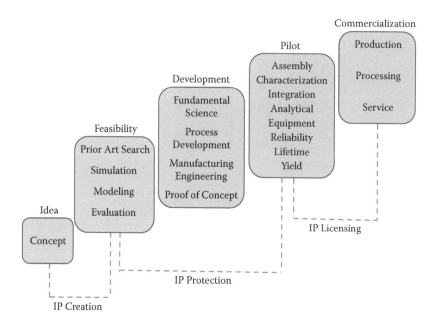

FIGURE 4.5
ANI's framework for IPR management.

ANI's framework for its integrated R&D, IPR management, and commercialization is described in their company website depicted in Figure 4.5.

Some of its licensing agreements are with Ishihara Chemical Company Ltd., a Japanese chemical company for conductive copper inks; Yonex Co. Ltd., a Japanese sporting goods company for nanocomposites; Multinational Power Transformers for hydrogen sensors; and Futaba, Inc. for display applications-electronic emissions. With Novus Partners, LLC it has a business relationship for digital signage networks, which is structured to earn future payments to ANI based on a percentage of the revenue of Novus Partners.

In September 2005, ANI signed a development contract with Yonex Co. Ltd. to develop nanocomposites to be used in sporting equipment. This agreement culminated in a license agreement in October 2008 that allowed Yonex to use ANI's technology in its tennis and badminton racquets. In 2010, ANI completed an additional agreement with Yonex to license this technology for use in its golf shafts. This development agreement served as the foundation of ANI's work in the CNT composites area. Yonex introduced these nanotechnology-incorporated golf shafts and badminton/tennis racquets in 2011.

ANI's work with Ishihara started with a feasibility study in early 2006 and culminated with a license agreement for copper inks and pastes in July 2009. Because of this partnership, ANI received over $2.5 million in research funding from Ishihara and a $1.5 million up-front license payment. Ishihara introduced its product in June 2011.

ANI has meticulously strategized their IP creation and commercialization by frequently performing funded research for both government entities and large corporations, which enables it to focus on customer-led areas and optimally utilize its resources to ensure speedy commercialization and maximize on profitability.

ANI generally spreads its risks by managing its projects in three phases, the first involving less than $100,000 possibly by self-funding for proving the feasibility and establishing laboratory-based proof of concept, followed by the second phase with outside funding of approximately $100,000 to $1,000,000 to establish the actual development of a product using the technology, or moving the technology from the laboratory to early stage development, and then a third phase, which is structured in conjunction with a strategic partner as a joint venture, subsidiary, or some other form of relationship.

Another key element in ANI's strategy is to design innovations and the associated IPR (especially patents) so that it is able to enforce its patent claims through strong demonstrative techniques. One such example in September 2011 is ANI's U.S. Patent No. 5,869,922, originally issued in the United States in 1999, claiming specific Raman spectrum signatures for all carbon films that have superior electron emission properties, which has also been issued in Japan (Japanese Patents No. 4786471 and No. 4786031), China, Korea, and Europe. The first claim of this group of patents states that: "A field emission device comprising a layer of carbon film on a substrate, wherein the carbon film has a UV Raman band in the range of 1578 cm^{-1} to 1620 cm^{-1}, wherein the UV Raman band has a full width at half maximum (FWHM) from 25 cm^{-1} to 165 cm^{-1}." The significance of this group of patents is that UV Raman spectra analysis can be easily executed at many universities and laboratories to identify if someone is using carbon films for electron emission that has the Raman spectra signature specified in ANI's patents. This provides ANI a powerful tool to monitor competitors and also defend and enforce the claims of its patents against any current or future violators.

ANI holds an extensive patent portfolio in the area of electron emission, and believes that this significant group of patents covers all carbon nanofilms, including carbon nanotubes used for electron emission applications. There is great interest in carbon nanotube innovation with companies small and large scrambling to integrate the technology into their newest electronics, x-ray equipment, lighting devices, LCD backlights, and field emission displays. Samsung, one of the world's largest electronics manufacturers, has already recognized the technology's importance and secured a license agreement in November 2010 as part of a package of approximately 150 patents that it licensed from ANI. While Samsung has a paid-up license to the technology, it is available for license to other companies on a nonexclusive basis. Success by Samsung in the introduction of a product using this technology clearly enhances the value of the patents and acts as a likely inducer to others to introduce products requiring a license.

This also demonstrates the business strategy of ANI on how to milk its dominating intellectual property from any area such as electron emission that is no longer part of its growth strategy or core business activities by synergizing the additional income from this portfolio to its earnings for years to come.

Another aspect that needs to be appreciated is the ability of ANI to handle patent-related litigations. Two such litigations are of significance. ANI had a dispute with Canon on an ANI patent and a dispute related to a licensing deal with Till Keesmann in 2000.

In May 2000, ANI licensed the rights to six carbon nanotube patents from Till Keesmann in exchange for a payment of $250,000 payable in shares of ANI's common stock. Under the terms of the agreement, ANI received the exclusive right to license these patents to others and was obligated to pay license fees equal to 50% of any royalties received by ANI specifically related to these patents. ANI was allowed to offset certain expenses, up to a maximum of $50,000 per year, against payments due under this agreement. The agreement also contained provisions related to minimum license fee payments. These minimum payments, totaling $1,000,000, were made and as a result, ANI received an irrevocable license to use these patents for the life of the patents.

ANI did not receive any license revenues and in 2008, the Keesmann agreement was amended as part of a litigation settlement. As part of that amendment, ANI agreed to return the licensing rights to Keesmann in May 2010 if minimum payments equal to $1 million were not paid to Keesmann by that date. No license revenue was received by that date and ANI made the conscious decision not to make the minimum payment and chose to allow the licensing rights to revert to Keesmann. Accordingly, the licensing rights for his specific patents reverted to Keesmann. However, ANI retains the irrevocable right to use the patents and is entitled to 50% of any royalties generated by those patents, up to a maximum of $1.2 million.

ANI attributes a lot of significance to confidentiality agreements and accordingly requires its employees, directors, consultants, outside scientific collaborators, sponsored researchers, and other advisors to execute confidentiality agreements upon the commencement of employment or consulting relationships with ANI. Further, all inventions conceived by individual employees also become exclusive property of ANI.

Since its inception, ANI has posted losses until December 31, 2009. However, it made a profit for the first time in 2010 and expects the profitable trend to continue through 2011.

In the case between Canon and ANI, which has been elaborated in Chapter 5, ANI dropped all charges against Canon after it received a decision against itself. This demonstrates the maturity of ANI to make strategic decisions on litigations on whether to continue or dispose of them based on techno-economic considerations.

4.7 Case Study 6: mPhase Technologies, Inc.

mPhase Technologies, Inc., a New Jersey corporation ("mPhase"), is a publicly held company founded in 1996. Prior to February 2004, the company was engaged in the telecommunication industry focusing on hardware/software solutions for telephone service providers for the delivery of voice, digital television, and high-speed Internet. The company discontinued these businesses in December 2007.

mPhase's R&D now focuses on the development of cutting-edge energy products based on "smart surfaces" using materials science engineering, nanotechnology science, and the principles of microfluidics and microelectromechanical systems (MEMS). The smart surface can be used for innovative self-cleaning applications, water purification/desalination, liquid filtration/separation, liquid and chemical sensors, controllable drug delivery, and environmental cleanup.

In February 2004, the company engaged the Bell Labs division of Lucent Technologies, Inc. in a $3.6 million project to develop basic smart nanobattery architecture, a new type of power cell energy storage device such as a reserve battery product with a virtually unlimited shelf-life prior to initial activation. This required the creation of a suspension, in droplet form, of liquid electrolyte on a "smart surface" or repellant such as silicon followed by an electronic impulse to trigger the process of "electrowetting" or collapse of the droplet and the mixing of the electrolyte, thereby providing a low level source of energy. In 2005, it created the first battery on chip using nano grass structure and in 2006, it produced the first generation battery lights LED.

mPhase's R&D is closely linked with leading institutions including Alcatel/Lucent Bell Labs from which it derives the latest advances in nanotechnology in the field of its business interest.

In July 2007, mPhase received a Small Business Technology Transfer (STTR) Program Phase I grant for $100,000 from the U.S. Army and in September 2008, was awarded $750,000 (net $500,000) Phase II STTR grant to continue battery development work for the SRAM project, which was renewed in 2009 for a second year to develop micro-array lithium battery for mission critical SRAM projects. In 2010, the company successfully completed the STTR by delivering a prototype smart nanobattery to the U.S. Army.

The company has also been working with the U.S. Army as part of a cooperative research and development agreement (CRADA) for the development of a lithium smart nanobattery. Under the Rutgers ESRG Collaborative Agreement in 2007, mPhase introduced the first version of the lithium smart nanobattery designed for portable electronics and microelectronic applications.

The company also developed a second product line "double barrel illuminator" designed by and co-branded with Porsche Design Studio (PDS) under a design agreement it signed with PDS in 2009. The product is an emergency flashlight designed primarily as an accessory product for automobiles and in 2011 has modeled its marketing over the Internet and sale through PDS's stores located in approximately 100 cities globally.

As a comprehensive strategy of retaining its lead position in high-technology products line, on July 28, 2011, mPhase signed a letter of intent to acquire Energy Innovative Products, Inc. (EIP), a privately held Nevada corporation. EIP is a developer of proprietary technology for reducing energy usage in refrigeration and cooling systems.

mPhase maintains a strong portfolio of know-how and patents either directly owned or licensed by the company from its strategic technology partners. A sample of the patent portfolio of granted and pending patent applications in the U.S. Patent Office are given below.

Granted Patents

Patent Title	Applicant (Ownership)
Tunable liquid microlens with lubrication-assisted electrowetting	Lucent Technologies
Method and apparatus for reducing friction between a fluid and a body	Lucent Technologies
Battery having a nanostructured electrode surface	Lucent Technologies
Method and apparatus for controlling the flow resistance of a fluid on nanostructured or microstructured surfaces	Lucent Technologies and University of Limerick
Nanostructured battery having end-of-life cells	Lucent Technologies and mPhase Technologies, Inc.

Pending Patent Applications

Patent Applications	Applicant (Ownership)
Structured membrane with controllable permeability	Lucent Technologies
Adjustable barrier for regulating flow of a fluid	mPhase Technologies, Inc. and Rutgers University
Battery system mPhase Technologies, Inc.	Rutgers University
Event-activated micro-control devices	mPhase Technologies, Inc.
Combined wetting/nonwetting element for low and high surface-tension liquids	mPhase Technologies, Inc.
Device for fluid spreading and transport	mPhase Technologies, Inc.
Reserve battery	mPhase Technologies, Inc.
Electrical device having a reserve battery activation system	mPhase Technologies, Inc.
Modular device	mPhase Technologies, Inc.
Reserve battery system	mPhase Technologies, Inc.

The sample patent portfolio clearly demonstrates the strategy of the company to use patent licensing arrangements with platform technology developers and then develop applications for which the patents are owned by the company. The company continually enhances its patent portfolio position in smart surface technology as is evident from the number of recent patent filings. An independent patent valuation of mPhase technology performed by First Principals, Inc., a technology appraisal and commercialization firm, estimates a minimum valuation of $40 million for its portfolio of patents and intellectual property.

4.8 Case Study 7: Bilcare Research, Pune, India

Bilcare Research is an innovation-led solutions provider that has partnered global pharmaceutical and healthcare industries to improve patient healthcare outcomes. Bilcare Research has a global footprint with modern R&D centers and manufacturing facilities in the United States, Europe, India, and Singapore.

In January 2008, Bilcare Singapore Pte Ltd, a wholly owned subsidiary of Bilcare Limited, bought 100% of Singular ID, engaged in research, development, and creation of micro- and nanotechnology-based novel products and was the provider of integrated high technology enterprise brand security system, for a consideration of Singapore $19.58 million. Singular ID has been integrated into Bilcare Technologies, a division of Bilcare Research, which works closely with its clients to tailor its technology to meet specific customer requirements. Singular ID had in-licensed two patent applications—PCT/SG2004/000216 (titled "A Method of Identifying an Object and a Tag Carrying Identification Information") and PCT/SG2005/000012 (titled "Identification Tag, Object Adapted to be Identified, and Related Methods, Devices and Systems")—from the Agency for Science Technology and Research (A*STAR), Singapore.

In the past few years, after acquiring Singular ID, the team at Bilcare Technologies has made path-breaking developments far beyond the technology covered by the basic patents that were in-licensed by Singular ID. Bilcare Technologies has now successfully developed novel proprietary nonclonable™ nano- and microstructured materials based tags, and specialty readers/scanners for security systems including anticounterfeit applications in brand protection and management. The unique magneto-optical signatures of the nonClonableID™ tags (randomly distributed micro- and nanoparticles that cannot be reproduced or duplicated) can be read and deciphered by specially developed Bilcare proprietary handheld/portable readers and associated software to authenticate them in real time in a foolproof integrated ICT system. This unique technology has been protected by a portfolio of issued patents and patent applications in various countries (PCT/SC2006/000160; PCT/SG2006/000159; PCT/SG2009/000056; PCT/SG2010/000259; PCT/ SG2011/000306). This technology significantly enhances security levels with an impregnable personal access control and identity management system that can be adapted for wide-ranging applications spanning security, anti-counterfeiting, etc. The epicenter of this system is the tamper-evident non-ClonableID™ nanotag capable of seamless and secure integration with any ICT system wherein the proprietary reading device scans the fingerprint and instantly communicates the encrypted information with a secure server through mobility platforms such as GPRS, 3G, or broadband to generate an instant complete authentication report on a mobile or computer using robust web enterprise–secured applications and data management at the back-end. The business model of Bilcare is represented in Figure 4.6.

Commercialization of this technology based on nanomaterials has involved close collaborative efforts of Bilcare with developers of the spe-cial reading devices, technology integrators to incorporate communication systems into the reading devices, and integrating the reading devices with

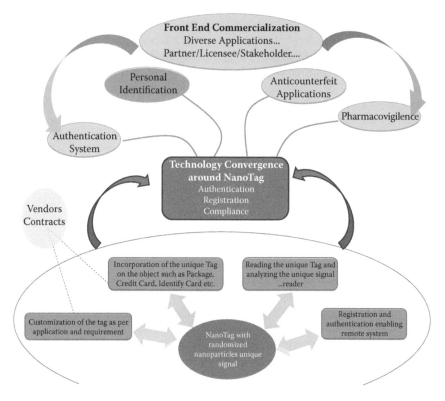

FIGURE 4.6
Bilcare business model.

mobile telephony systems including developers of secured enterprise management systems. Another set of collaborators has been those having expertise to incorporate the nonClonableID™ nanotags into diverse substrates in a robust manner to withstand harsh operating ecosystems and platforms where the nonClonableID™ nanotag retains its unique signature and tamper evident characteristics.

The first signs of commercial applications of Bilcare Technology's non-ClonableID™ nanotag has been the highly secure identity cards adopted by the Delhi Police in India for real-time authentication of the individual police officer and centralize the duty-planning rooster for planning and monitoring. The company is now working closely with the pharmaceutical industry, agrochemical industry, and high value components and luxury goods offering these sectors structured anticounterfeit, e-pedigree, and secure track-and-trace increased visibility across the supply chain, patient compliance, and clinicom solutions including tackling issues of diversion and theft in the supply chain.

Such an operating business model offering specialized products and services demands focused understanding of the clients' application needs coupled with nimble footedness to flexibly and cooperatively network in real time with partners having diverse expertise. The business operations in such a model based on next generation technologies has to continually confluence strategic management of innovations, IPR, confidentiality, benefit sharing arrangement for profitable cash flows for the operating partners, and regulatory compliance, including risks and liabilities.

4.9 Case Study 8: HyCa Technologies Pvt Ltd., Mumbai, India

HyCa Technologies is an emerging cleantech company started by Professor A. B. Pandit of the University Institute of Chemical Technology, Mumbai and Anjan Mukherjee, a marine engineer and entrepreneur, exploiting energy dissipated by collapsing cavitation bubbles to modulate physical, chemical, and biological processes.

The phenomena of cavitation involving nucleation, growth, and the subsequent collapse of a formed cavity is a means of delivering required transformational energy (required for chemical, physical, or biological transformation) on a nanoscale most efficiently. The cavity (vapor, gaseous, or mixed), which starts at a nanoscale nuclei generation (1 to 2 nm), grows to a few hundred microns and then collapses again to practically nanoscale generating a nanoscale dimension spherical pressure or a shock wave and can release this collected energy from sonic irradiation or fluid kinetic energy in the form of a concentrated packet. The intense and extreme local energy dissipation on a nanoscale can create a nanoenvironment, which creates nanoparticles, in a top-down approach, or restrict the growth of the particle in a bottom-up approach. This is due to enhanced transport conditions created due to microstreaming generated by collapsing or oscillating cavities.

Hyca Technologies, using its patent-pending core innovation the HyCator™ design algorithm that enables its reactors to create targeted cavitation bubbles required for the specific process needs, exploits tailored cavitation to harness energy dissipated by collapsing cavitation bubbles to accelerate chemical reactions, break down complex molecules, and perform uniform mixing to develop environmentally friendly and cost-effective applications ranging from effluent treatment, biofouling prevention, and ballast water treatment to chemical processes and cooling towers, across industries ranging from effluent treatment to petrochemicals involving flowing fluids or slurries.

HyCa's business model has been to develop a product, do the initial consultative sales process, and then license its technology to a business partner. The present focus of the company is the development of HyCator™ reactors for effluent treatment, chemical processes, and cooling tower and ship ballast water treatment. The authors have represented the business model in Figure 4.7.

The company has received angel funding from Godrej Industries Ltd and a loan under the SPREAD Scheme from World Bank administered by ICICI. It is generating its cash flow from several reputed customers such as Rallis, Unilever, Godrej, Atul, Marico Ltd., and the Tata Group through offerings of its HyCator for various applications. HyCa's plan now is to develop its manufacturing facility, expand the team, file additional patent applications to create a strong patents portfolio, seed the identified segments, and team up with global leading channel partners to bring the products to market.

4.10 Case Study 9: Consolidation through Sequential Merger of Carbon Nanotechnologies, Inc. (CNI), Nanopolaris, Unidym, and Nanoconduction, and Acquisition of Unidym by Wisepower Co. Ltd.

Carbon Nanotechnologies Incorporated (CNI) was founded in 2000 based on the groundbreaking work on carbon nanotubes (CNT) by Dr. Richard Smalley's team. Dr. Smalley's pioneering work led to the development of a portfolio of more than 100 patents (including 54 issued U.S. patents) owned by CNI or exclusively licensed to CNI by Rice University. Since its inception, CNI provided bulk CNT materials to customers and won research grants

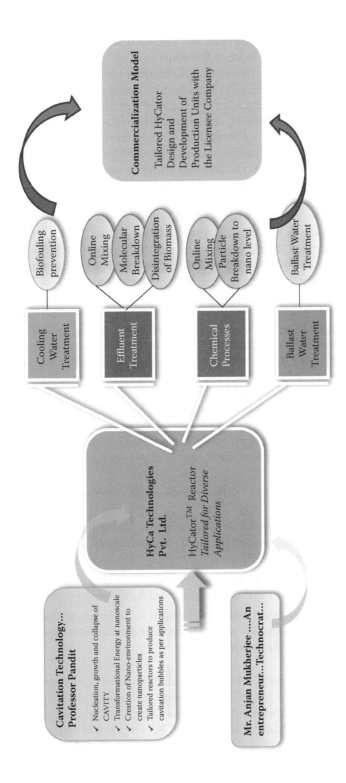

FIGURE 4.7
HyCa Technologies business model.

from government agencies such as the National Institute of Science and Technology (NIST) and the State of Missouri.

In 2005, NanoPolaris was founded and obtained exclusive licenses to university CNT patents. In 2006, CNI launched its initial commercial products. Further, In 2006, NanoPolaris acquired the assets of an early stage company called Unidym, a subsidiary of Arrowhead Research Corporation. At the time of the acquisition, NanoPolaris had already consolidated certain intellectual property related to carbon nanomaterials. Through this acquisition, NanoPolaris gained access to the former Unidym's substantial expertise and intellectual property in CNT films. After its purchase of Unidym's assets in June 2006, NanoPolaris changed its name to Unidym.

In 2007, Unidym merged with CNI. This pooling of investment capital, alignment of strategy, and integration of materials and device production reflect moves that dictate the imperatives for the successful commercialization of these nanomaterials. In 2007–2008, it obtained $14.4 million Series C financing, including participation by Battelle Memorial Institute, Entegris, and Tokyo Electron.

In 2008, Unidym got involved in a joint development with Samsung Electronics to produce the world's first CNT-based active matrix electrophoretic display (EPD) e-paper, which was demonstrated at the Society for Information Display (SID) International Symposium. In 2008, Unidym demonstrated the world's first full-color active matrix LCD made with CNT components using production capable and scalable manufacturing processes at the SID International Symposium. Further, in August 2008, Unidym acquired Nanoconduction, Inc.

Tokyo Electron Venture Capital and NanoGram Corporation analyzed the Unidym/CNI 2007 merger and drew four central conclusions[6]:

1. The need for nanotech companies to focus on real products and revenues.
2. Integration of nanomaterial suppliers with companies focused on developing products incorporating nanomaterials is necessary to achieving commercial success.
3. The need to consolidate the fragmented patent landscape is important to facilitate the development of the industry as a whole.
4. The decision of the combined company to focus on electronics applications reveals that CNTs and other nanomaterials have their greatest impact in the near future.

At the time of the merger, CNI had a patent portfolio of more than 100 patent filings in diverse aspects of CNTs, but despite that strong patents position, it decided to merge with Unidym in an aggressive, cross-industry partnership strategy to foster the commercialization of CNTs in a broad range of applications. With the addition of the experience and strengths of CNI, Unidym's current patent portfolio contains more than 200 foreign and domestic patents and patent applications, including more than 90 issued patents.

Unidym's foundational patent portfolio covers nearly every aspect of CNT manufacturing and processing covering CNT compositions (high-purity CNTs, derivatized CNTs, electrical conductors containing CNTs, CNT ropes, CNT composites, CNT dispersions, CNT arrays, and CNT fibers); production (synthesis, purification, dispersion, production of CNTs using methods, such as electric arc, laser vaporization, chemical vapor deposition, and predominantly gas phase processes); and application-enabling technologies (postsynthesis treatment of CNTs to purify, disperse, functionalize, and otherwise alter them). Unidym's portfolio includes claims related to the modification and manipulation of CNTs, which range from cutting CNTs, opening the ends of CNTs, purifying CNTs, end-derivatizing CNTs, side-wall derivatizing CNTs, noncovalently derivatized CNTs, doping CNTs, aligning CNTs to dielectric materials containing CNTs, and articles of manufacture, which include CNT arrays, transparent electrodes, thin film transistors, fuel cell components, super capacitor components, battery components, memory, electrochemical devices, field emitters, photo-emitters, optoelectronic devices, catalyst supports, electromagnetic shielding, electromechanical transducers, sensors, conductive composites, structural composites, medical devices, and drug delivery.

Through the merger of Unidym with Nanoconduction, Unidym had access to Nanoconduction's patent portfolio, which supplemented Unidym's existing patent portfolio and provided Unidym with additional opportunities to out-license and leverage its technology. In addition, through the merger, Unidym gained access to facilities and equipment of Nanoconduction that would be used in Unidym's ongoing research and development activities.

With its licensing program to commercialize products outside of Unidym's product development efforts, technical expertise, and manufacturing facilities, Unidym enables partners to rapidly develop CNT solutions for their specific applications. Companies that might benefit from a license to Unidym's patent portfolio and know-how include

- A company seeking to manufacture nanotube-based products that would like to synthesize CNTs as part of its manufacturing process
- A company seeking to manufacture nanotube-based products that would like to purchase the CNTs from a supplier other than Unidym
- A company interested in supplying CNTs

In January 2010, Unidym was acquired by Wisepower Co., Ltd., a publicly traded, Seoul, Korea-based electronics company that is a leading supplier of Li-polymer batteries for mobile appliances in Asia. The company has developed high-quality LED packages and solid-state lighting and has a strong portfolio of products that include wireless charging systems for electronic applications. Wisepower also has a wide customer base that includes cellular phone manufacturers such as LG Electronics, Pantech, and KTFT, and has deep relationships with many potential Unidym customers. Wisepower's experience in the growing electronics business found a strategic fit with Unidym's leadership in CNT technology for electronics applications. The combined strength positions the company to exploit the mutual expertise and facilities to develop manufacturing scale-up and customer acquisition as it moves into the market adoption phase of its business.

Upfront consideration to Unidym shareholders consists of Wisepower stock and convertible bonds valued at US$5 million with certain restrictions as to timing of stock sales. Unidym shareholders are also entitled to cash earn-out payments up to US$140 million based on cumulative sales and licensing milestones. Finally, Unidym shareholders will receive 40% of licensing revenue from a set of patents that have been generating increasing quarterly royalties in addition to the potential $140 million earn-out.

This is an excellent example that demonstrates that as early as 2005 Arrowhead saw an opportunity in CNTs and started the company, which had the potential to significantly impact multiple large and diverse industries. At the time Unidym was launched, the CNT market was highly fragmented with key patents dispersed across multiple owners and there was no clear industry leader. Unidym, as has been elaborated, licensed technology from several universities and acquired three prominent CNT companies, including Carbon Nanotechnologies, Inc., the pioneering company in high-performance CNTs, and has emerged as a leader in the development of innovative CNT-enabled products for the electronics industry. In this process, Unidym assembled a strong and diverse patent portfolio that strategically covers high-performance CNT manufacturing and processing, as well as multiple product applications. The acquisition of Unidym by Wisepower in 2010 demonstrates the need for a strategic partner that can bring about a horizontal integration of the two businesses to create the appropriate organizational mix to accelerate targeted market penetration for speedy revenue generation.

4.11 Case Study 10: Innovalight and DuPont Merger

Innovalight, founded in 2004 in St. Paul, Minnesota as a technology incubator, focused on material science for solid-state lighting applications. It filed its first patent in January 2006 and demonstrated the first solar cell results with silicon ink. In April 2006, it also raised $7.5 million in new financing capital led by investors from New York, the Harris & Harris Group Ltd.

In October, $28 million was raised in new financing capital led by investors from Norway, Convexa Capital. In November 2007, Innovalight's patent portfolio had 30 patents issued/pending in materials and devices. In January 2008, it moved to Silicon Valley headquarters in Sunnyvale, California and in September 2008 Innovalight was awarded a $3 million grant from the U.S. Department of Energy.

The year 2009 was a watershed year for Innovalight when the first solar panel was produced using Innovalight technology. In May 2009, printable silicon ink was developed and 18% conversion efficiency for 156-mm silicon ink inkjet-based solar cells was achieved in June. Further, in July, JA Solar produced the first silicon ink solar cell at a customer site. In September, Motech Industries signed an evaluation agreement with Innovalight followed by Chevron installing test modules based on Innovalight, silicon ink in Bakersfield, California in November and JA Solar signing the first multiyear commercial contract with Innovalight in December.

In January 2010, Innovalight secured $18 million in new financing, led by Singapore-based investors, EDB and Vertex Management. In June, JA Solar signed a development contract with Innovalight followed by the signing of a development contract with Yingli (China). In September 2010, Fraunhofer Institute confirmed 19% conversion efficiency for silicon ink–based 156-mm solar cells followed by Hanwha SolarOne signing a commercial contract with Innovalight in October and a commercial contract with JA Solar in December. By the end of 2010, Innovalight had a patent portfolio of 65 patents issued/pending.

In first quarter of 2011, Innovalight commissioned its second-generation silicon ink production line at the Sunnyvale site. In May 2011, Innovalight was awarded $3.4 million as part of the U.S. Department of Energy SunShot development program. JA Solar ramped up production of SECIUM™ solar cells with silicon ink from Innovalight.

In July 2011, DuPont acquired Innovalight. The acquisition further strengthened DuPont's position as a clear leader in materials for the solar energy market, enabling a broader and more integrated photovoltaic (PV) materials and technology offering from DuPont with technology that improves cell efficiency for expedited adoption for more efficient and affordable solar energy. Silicon inks used in conjunction with DuPont™ Solamet® photovoltaic metallization pastes boosted the amount of electricity produced from solar energy, enabling the production of superior selective emitter solar cells. According to industry estimates, selective emitter technology could represent 13% of crystalline silicon solar cell production by 2013 and up to 38% by 2020.

This merger in nanotechnology vindicates those who believed offerings complementary to one another and together in solar devices, silicon technology, and selective emitter technology by Innovalight, and DuPont's expertise in materials science, manufacturing capabilities, and global market access broaden and accelerate their ability to meet customer needs and address the emerging challenges with their continued innovations. DuPont Innovalight will continue as a wholly owned subsidiary of the parent company DuPont.

4.12 Invitrogen (Presently Life Technologies Corporation)

Founded in 1987, Invitrogen, headquartered in Carlsbad, California, conducted business in more than 70 countries. In 1989, Invitrogen was incorporated and in 1999, it completed its public offerings of stock.

In late 1989, founding partner Bill McConnell left the organization to form his own company, called McConnell Research. McConnell's departure forced one of the co-founders of Invitrogen to use most of Invitrogen's capital to purchase the departing founder's interest in the company. The leveraged buyout left the fledgling company in a precarious position. Without the collateral required to secure a bank loan, Invitrogen searched for another source of financing and found aid from the California Export Finance Office (CEFO). CEFO was a division of the World Trade Commission, an organization chartered to help small and mid-sized California businesses that needed capital to expand into foreign markets. The aid provided by CEFO resolved Invitrogen's financial crisis, enabling the young company to gain its footing and expand internationally. As of 2004, the company employed approximately 4500 scientists and professionals globally and had revenues of more than US$1 billion.

The company went on a strategic acquisition drive to bolster its capabilities and to create an invincible patents portfolio to operate on a robust business platform in the field of biomolecular labeling. A few key transactional milestones of Invitrogen are indicated in Table 4.1.

The company's patented TOPO cloning technology dramatically accelerated the cloning process, reducing the time it took for one step in the process from 12 hours to 5 minutes. With the success of its patented cloning technology and the litany of other Invitrogen products, the company was able to record respectable growth throughout the 1990s, moving well beyond the $4 million recorded at the beginning of the decade. By the company's 10th anniversary, it had eclipsed the $50 million in sales mark, generating $55.3 million in revenues.

Much of Invitrogen's subsequent financial growth has been fueled via the aggressive acquisitions indicated in Table 4.1.

The acquisition of Quantum Dot Corporation and BioPixel® by Invitrogen was completed in October 2005. Invitrogen, with its molecular probes, was then a leading life science company providing innovative labeling and detection technologies to support disease research. Quantum Dot Corporation offered novel solutions for biomolecular labeling and detection that employ

TABLE 4.1

Transactional Milestones

Dates	Transactions by Invitrogen
August 1999	Merger with San Diego–based NOVEX, a developer of products used for gene and protein analysis, in a $52 million deal.
December 1999	Research Genetics, Inc., Huntsville, AL, for $139.2 million in stock.
June 2000	$15.1 million acquisition of Ethrog Biotechnologies Ltd. of Israel, which had developed and patented a system for the electrophoretic separation of macromolecules. This was followed by a cash and stock merger valued at $1.9 billion.
July 2000	Dexter Corp, a chemical maker, and Life Technologies, Inc., which makes biological material for genetic research, for $1.9 billion in cash and stock. Dexter has 75% stake in Life Technologies.
September 2000	Completes its takeover of chemical maker Dexter.
March 2001	Invitrogen agrees to sell division housed on 240,000-square-foot site to Human Genome Sciences, Inc. for $55 million.
October 2002	$42 million agreement to acquire Informax, which developed software that helped to design, manage, and interpret research kits for gene identification and cloning.
July 2003	Acquires Molecular Probes, Inc. for about $325 million to add drug-discovery products.
December 2003	Acquires BioReliance Corp. for $430 million to expand its production ability for biotechnology customers.
February 2005	Agrees to acquire closely held Dynal Biotech for $381.6 million to gain technology for cell research to speed development of products.
July 2005	US$130 million acquisition of BioSource International with expertise in functional proteomics. This deal augmented Invitrogen's growing collection of protein and primary antibody products gained through its earlier acquisitions of Zymed Laboratories and Caltag Laboratories.
October 2005	Announced the acquisitions of Quantum Dot Corporation and the BioPixels® business unit of BioCrystal, Ltd. and the early closing of Biosource International, Inc.
June 2008	Agrees to $6.7 billion in cash and stock to buy Applied Biosystems, the company that provided most of the equipment for Human Genome Project and created a supplier of machines and materials for use in academia, pharmaceutical industry, R&D laboratories, with about $3.5 billion annual sales. The overall company is called Life Technologies.
October 28, 2008	Buys VisiGen for $20 million to bolster its third generation patent from using the expertise and IP of the budding developer of a real-time, single-molecule sequencing-by-synthesis technology to become the leader in the new genomics era.
Mid-2010	Acquisition of computer chip DNA sequencing company Ion Torrent Systems.

Quantum Dot (Qdot®) semiconductor nanocrystals. The company also held the broadest intellectual property portfolio in the life science industry for semiconductor nanocrystals with more than 160 patents and applications, and had built a significant customer base that was using this latest labeling and detection technology. BioPixels® provided novel specially coated fluorescent nanocrystals and metal alloys for applications in multicolor labeling, sorting, and imaging of cells, lateral flow immunoassays, and fluorescent inks, and represented a promising technology for the development of automated assays of complex biological samples. The combination of the three companies allowed the creation of smaller, brighter, lower toxicity particles that do not blink, with an enhanced patent portfolio.

At around the same time, Invitrogen also announced an agreement with Georgia Tech Research Corporation to exclusively license novel "nanocluster" technology. Taken together, the combination of these acquisitions and licenses provided Invitrogen with a significant intellectual property position and robust platform for cutting-edge product development. Further, the added capability enabled Invitrogen to create new innovative products based on advanced inorganic materials science for molecular detection that enable life science researchers to better visualize and understand cellular processes, molecular interactions in proteomics, genomics, gene expression, and imaging and other factors essential to diagnosing and treating disease. Terms of the acquisitions and license were not disclosed.

In 2008, Invitrogen virtually doubled its size with the purchase of a biotech instrumentation company called Applied Biosystems, maker of DNA sequencing and polymerase chain reaction (PCR) machines and reagents. The company then renamed the overall organization Life Technologies. The Invitrogen brand and most of the brands acquired still exist on product packaging, although the overall company is called Life Technologies.

In summer 2010, the company acquired the computer chip DNA sequencing company Ion Torrent Systems. In June 2010, Evident Technologies, Inc. admitted infringing three patents and agreed to an injunction as part of a settlement in a case brought by Invitrogen Corporation (now known as Life Technologies Corp.) over quantum dot semiconductors. The parties stipulated that Evident had infringed and induced infringement of the patent-in-suit, and that the patents' claims were valid in all. Judge Leonard Davis of the U.S. District Court for the Eastern District of Texas signed off on a consent order and permanent injunction in the case.

This is a story of a start-up company that has experienced all the steps as an enterprise from its birth to growth to its emergence as a successful and profitable enterprise taking into consideration a business model to support targeted technology development, commercialization, procurement of funds from diverse sources, ensure growth through planned mergers and acquisitions, timely and appropriate exit from unprofitable activities, enforcement of its intellectual property rights to establish dominance, network with stakeholders and clients, and creation of a sustainable business.

Through this history of acquisitions and continued product research and development, Invitrogen/Life Technologies now has over 50,000 products.

4.13 Strategic Investing and the Role of Venture Capital

Venture capital providers have been sources of short- and long-term funding generally with the investor buying a share of the company, and the amount that is paid for this shareholding flowing into the company as new equity. In several cases, in addition to such financing, the venture capital investors value add to the new enterprise with their commercialization experience, business skills, and know-how to facilitate or manage the process of product development, market creation, and product launch to achieve profitable growth. This helps the investor's portfolio companies to develop more robustly so that the individual company as well as the shareholding of the investor increases in value. In addition, the company has a further benefit: If it should fail, the founders do not have any personal liability toward the other investors.

Angel investors, who generally are individuals or a group of individuals, invest their money in promising start-ups to see them take off usually in exchange for convertible debt or ownership equity and then appropriately exit hopefully with profit. Angel investors have limited ability to take risks. On the other hand, banks generally lend against some security and seldom are prepared to take high risks. In contrast, venture capital investors are prepared to take reasonable risk as they create their "security" provisions to minimize losses if any to their shareholders/investors. The "due diligence" process undertaken by venture capital investors is fairly rigorous as they evaluate the technology, patent portfolio, target market, technical and professional expertise of the promoters, management track record, valuation of the company and its intellectual and physical assets, etc. before making their investments and in exchange deriving benefits from the company through diverse options such as taking stake in the company through equity or becoming part of the management team.

The "due diligence" is extensive, as the venture capital investor has to assess the chance of success so that it achieves its key goal of selling its shareholding in the company, maximizing its earnings, and sharing the profit with its shareholders.

We now illustrate the role and functioning of a typical venture capital investor, Nanostart AG, in nanotechnology.

4.13.1 Case Study: Nanostart AG

Nanostart is a leading nanotechnology investment company that has strategically invested in young, up-and-coming companies that seek to commercialize highly innovative and promising nanotechnology-based products or processes. It maintains a global portfolio of companies including research centers in Germany, Silicon Valley, and Singapore. Nanostart is currently invested in nine portfolio companies, which are managed out of Nanostart's offices in Frankfurt, Berlin, and Singapore. Its present portfolio of nine companies with its share holding indicated in the parentheses is Magforce (61.2%), HolMenkol (50%), Namos (26%), ItN/Nanovation (19.3%), BioMers (24%), Lumiphore (19.6%), Microlight Sensors (32%), MINT (18.2%), and Nanosys (0.3%). In addition, it manages two funds, namely, Nanostart Singapore Early Stage Venture Fund I and Kama Fund First in Russia.

As a venture capital provider, it chooses to invest in appropriate phases in a company's life cycle either directly by Nanostart AG or through a fund structure to reasonably ensure rapid growth profitability.

Nanostart also value adds to its portfolio companies with its advisory support, management experience, and commercialization know-how, and further links its companies with Nanostart's global network of contacts in business, universities, research institutions, and government agencies, as well as among investors and capital market experts.

Every portfolio investment is made with a view toward a successful exit. Nanostart's objective is therefore to sell its holdings when the time is right to a commercial buyer for a substantial profit.

4.14 Bonds

The Singapore Early Stage Venture Fund I is a venture capital fund of Nanostart that invests in young nanotechnology companies in Singapore to support them in developing and commercializing their products especially where nano-enabled technologies have a strong impact on industry development. It finances entrepreneurs in life sciences, medicine, cleantech, and new materials in multiple financing rounds as primarily the lead investor in the first round with co-investments thereafter. The fund is a joint project of Nanostart AG and the Singaporean government. The government represented by the National Research Foundation (NRF) committed 10 million SGD (approximately 5 million euro) to the fund. Nanostart invested the same amount of money. The fund is managed by Nanostart Asia Pte Ltd, Nanostart's 100% subsidiary in Singapore that was founded in 2008.

Kama Fund First is a nanotechnology fund that was started with the Russian company RUSNANO and the governor of the administrative district Perm region. The closed-end venture capital fund has a total volume of €50 million and a term of 10 years. The aim of the fund is to invest in promising projects and companies in the field of nanotechnology. Nanostart views this fund as the ideal market entry in Russia.

A brief description of the portfolio companies of Nanostart illustrates the investment strategy of the group.

4.14.1 MagForce AG

Magforce AG is a leader in the area of nanotechnology-based cancer treatment, and the first company worldwide to receive European approval for a medical product using nanoparticles. The company's NanoTherm® therapy is a novel therapy for the treatment of solid tumors. MagForce is preparing for its product launch in Germany. A letter of intent was signed with Delrus, a leading Russian medical product distribution company, for market entry in Russia. Titles of some of the patents from Magforce's patent portfolio are given next as illustration of the strong proprietary support given to Nano-Activator™.

- Method for encapsulating therapeutic substances in cells
- Implantable products comprising nanoparticles
- Temperature-dependent activation of catalytic nucleic acids for controlled active substance release
- Varying magnetic field application device for heating magnetic or magnetizable substances in biological tissue
- Magnetic transducers
- Nanoparticle active ingredients conjugates
- Medical preparation for treating arthrosis, arthritis, and other rheumatic joint diseases
- Stent for keeping tubular structures open
- Method for cultivating cancer cells from human tissue and device for preparing tissue samples

4.14.2 BioMers Pte Ltd.

The Singapore-based medical technology company founded as a spin-off from the National University of Singapore (NUS) in 2005 with a branch office in the United States is developing innovative nanotechnological solutions for the orthodontic market. In 2010, it commercialized a fully transparent brace system called SimpliClear™ based on its self-developed nano plastic wire. BioMers has the exclusive worldwide licensing rights. The capital invested by Nanostart is primarily used to strengthen sales and marketing resources and to further expand the existing network of international distribution partners. The priority patent application numbers US20070839393 20070815 also published as WO2009022986(A3), US2009047614(A1), EP2178457(A2), and CA2694855(A1), titled Dental Retainer by Mervyn Fathianathan, Renuga Gopal, A.P.A.Bindu

Saran, Shanti Choudhury Rachel, Teo Chieh Yin Karen, Aliphtrias George, and assigned to BioMers support the technology of SimpliClear™.

4.14.3 Holmenkol AG

Holmenkol AG is known as the world's oldest producer of ski wax. Since 2002 it has specialized in nanotech-based surface coatings for a wide range of sporting applications where these surfaces come into contact with water in its various forms. In addition to high-performance ski waxes, the product range protected by 31 patents/patent applications (as searched in Espacenet) includes waterproofing, detergents, coatings, and polishes for a wide variety of outdoor, cycling, and aquatic sports.

Holmenkol's patents/patent application titles include the following:

- Sliding aid modifier for improving the tribological sliding properties of winter sports equipment on snow; lubricant for sports device
- Antiadhering agent, useful for coating transparent substrates such as ski glasses or protection helmet visor; comprises an optionally saturated carbonic acid amide
- Use of a composition containing fluorine copolymer with structural units of defined formula useful as a lubricant for winter sports equipment, especially skis and snowboards
- Device for determining the most suitable ski wax under certain conditions
- Paint for nonslip coatings
- Halterungen fuer daemmplatten
- Method of producing heat-insulated plaster facades
- Vorrichtung zum trocknen von auf einen parkettboden an der verlegungsstelle aufgetragenen lack
- Verfahren zum versiegeln von parkettboeden und vorrichtung zur durchfuehrung des verfahrens tischtennisplatte
- Aussenseitige waermedaemmung fuer gebaeudefassaden mit integriertem waermeaustauscher fuer waermepumpen
- Neuartiger verbundwerkstoff
- Absitzstabile suspensionen anorganischer erdalkaliverbindungen
- Verfahren zur herstellung derselben und verwendung der suspensionen zur abwasserneutralisation und abwasserreinigung

4.14.4 ItN Nanovation AG

ItN Nanovation AG was formed in 2000 and went public in 2006. It is one of Germany's leading nanotechnology companies, developing innovative ceramic products, such as filtration systems and protective and catalytic coatings in various industries from baking ovens to aluminum foundries to coal-fired power plants including large industrial customers. The company's success is founded on an extensive product portfolio, combined with comprehensive development and application expertise. In October 2011, ItN Nanovation announced a major commercialization success with the finalization of the agreements on a joint venture in Saudi Arabia for the production and sale of ceramic flat membranes (CFM Systems) throughout the Middle East and Northern Africa region. ItN Nanovation has a portfolio of over 250 patents/patent applications covering coatings, ceramics, metal particles, self-cleaning layers, colloidal systems, casting and molding production, ceramic filter element for water purification, filter devices, functional ceramics, antimicrobial polymeric coating compositions, unagglomerated core-shell particles, optical waveguides, tunable resonators, phase shifters, and nanophotonic directional coupler.

4.14.5 Namos GmbH

Namos GmbH (formerly operating under the name BoneMaster GmbH) founded in 1998 is a successful spin-off from the Technical University Dresden (Germany). Namos engages in R&D in the bio-nanotechnology field with the objective of developing defined surfaces with the help of biological molecules. Namos has successfully developed a particular bioactive coating for bone implants. Meanwhile, one of the largest global implant manufacturers has brought this coating to market.

A breakthrough technology from Namos GmbH involves a new process using an "intelligent," bio-nanotechnology-based coating of the ceramic substrate material used in automotive catalytic converters. This technology has the potential to save large quantities of precious metals in the production of automotive catalytic converters. Namos is focused on commercializing this proprietary technology. The potential markets for the new technology include not only the automotive sector but also the chemical and petrochemical industries. The supplier chain for automotive catalytic converters begins with the producers of ceramic powders and substrates, who deliver these to coaters (precious metal processors), onward to exhaust system providers, and finally to the major automobile producers.

A few patent/patent applications titles are given as follows:

- Semiconductor–nanoparticle functionalizing method for marking and detecting biological molecules in Western blot; involves deactivating free coupling group by simultaneous or subsequent addition of nonactive biomolecule

- Process for producing finely divided, high-surface-area materials coated with inorganic nanoparticles, and also use thereof

- Katalysator mit einer metallischen Nanostruktur auf Basis metallisierter Biotemplate sowie deren Verwendung
- Metallische Nanostruktur auf der Basis hochgeordneter Proteine sowie Verfahren zu deren Herstellung

4.14.6 Lumiphore, Inc.

Lumiphore Inc., in Redwood City, California, is a spin-off from the University of California, Berkeley based on the research of Professor Kenneth N. Raymond and his co-workers.

With its patented technology of Lumi4® complexes, the company has created a new generation of drug testing through which numerous medications or illegal drugs can be detected at the same time. Major global life sciences corporations such as Cisbio and Brahms AG have integrated the Lumi4® technology into their high-throughput screening (HTS) systems, used extensively in drug discovery. The technology also facilitates the rapid diagnosis of certain diseases, and offers promising approaches in the field of cancer treatment.

Titles of a few patents owned by Lumiphore are given as an illustration.

- Radiopharmaceutical complexes
- Macrocyclic hopo chelators
- R multicolor time resolved fluorophores based on macrocyclic lanthanide complexes

4.14.7 Microlight Sensors Pte Ltd

Microlight Sensors Pte Ltd in Singapore has designed, developed, and assembled fully integrated optical sensors and scanning systems for specialized applications in the homeland security and commercial vision systems markets. Microlight has developed strong and active partnerships with institutions and customers worldwide for its proprietary sensor in unique high-end optical systems.

At present, with significant advantage over its competitors, Microlight is serving a gap in the Asia Pacific Homeland Security market by providing completely integrated open-platform optical systems with after sales support. Eventually, many of Microlight's optical sensor technologies will transition to the commercial and industrial application markets, representing further growth potential for the company.

4.14.8 Membrane Instruments Technology Pte Ltd (MINT)

MINT, a cleantech company in Singapore, based on a technology originally developed at Singapore's Nanyang Technological University, has advanced its development of the next-generation, fully automated sensor systems that enable water treatment plants to raise their water quality and process efficiency to unprecedented levels. With its patented technology, MINT is able to produce uniquely sensitive sensors for quality control in water treatment plants providing major savings in the cost and time associated with water treatment monitoring and enables immediate reaction in case of malfunction. The company is now an active player in water treatment markets.

4.14.9 Nanosys, Inc.

Nanosys, based in Palo Alto, California, has a portfolio of over 670 patents and is one of the world's leading nanotechnology companies in the field of electronic nanostructures having strategically positioned itself in growth markets of IT/electronics and cleantech to provide customized products and applications on a virtually off-the-shelf basis and via its partnerships with leading global corporations that include Sharp and Intel, as well as the U.S. Department of Energy. Nanosys's technology serves as the essential basis for constantly better and cheaper LEDs, more compact storage devices, fuel cells to power mobile phones, and flexible circuit boards and solar cells, to name just a few applications. In the third quarter of 2011, Nanostart's portfolio company Nanosys, Inc. successfully launched its quantum dot enhancement film (QDEF) technology.

The business model of Nanostart AG illustrates the strategy and functioning of a comprehensive venture capital investor and its role in innovative value creation in the development and commercialization of evolving/disruptive technologies that require a vision to look and feel the future, assessment of a business opportunity, the infusing of funds with appropriate due diligence, coordinated efforts of continual technology development and innovations, deriving synergies from diverse sources, converting product concepts into marketable products, creating and effectively managing the intellectual property rights, networking relationships and partnerships, developing markets and sustainably servicing them, all with the goal of establishing a dynamic portfolio of profitable companies and profitably transacting the assets (both intellectual and physical) with a targeted exit at appropriate stages thereby ensuring sustainability of the venture capital investor with simultaneous value realization for its shareholders.

There are several venture capital investors, some operating globally and others operating within defined territorial limits, that are constantly in search of promising ideas, entrepreneurs, and business concepts in which to invest thereby effectively transforming dreams into reality. A list of venture capital investors (although not exhaustive) is given in Table 4.2.

TABLE 4.2

Venture Capital Providers and Angels in the Nanotechnology Sector

Name of Organization	Services
MP Systems	Silicon Valley consulting firm specializing in identifying and evaluating very early-stage companies (pre-VC) for investment and partnerships in nanotechnology, organic light emitting diodes, semiconductor equipment, and processes.
NanoBioNexus	Facilitation of partnering and investment opportunities in nanobiotechnology in the San Diego region by providing visibility into scientific breakthroughs in the nanobiology field through educational and networking programs and market research.
Silicon Valley Nano Ventures	Nano businesses development services including fundraising, sales, joint ventures, strategic partners, senior technical and executive recruiting since 2001.
The Strategic Synergy Group	Services team and business building in nanotech domain, executive and senior recruiting, technical and marketing consulting, and fundraising via its investor network including development, funding, manufacturing, and marketing in Asia-Pacific, through joint ventures and strategic alliances.
NanoSIG	Angel, VC, and corporate investors access to companies, IP, people, products, and technologies (and vice versa), by hosting nano funding and partnering opportunities in Silicon Valley and by partnering with affiliates globally.
Harris & Harris Group (Proposes name change to Small-Technology Venture Capital, Inc.)	Investment in private development stage or startup companies and in the development of new technologies in diverse industry segments including nanotechnology.
Axiom Capital Management	Builds and manages wealth for high-net-worth individuals and institutions.
Molecular Manufacturing Enterprises Incorporated (MMEI)	Seed capital, advice, contacts, and other support services to researchers and business people developing art of molecular nanotechnology.
OVP Venture Partners	Provides initial investment capital of $1 to $5 million to partner with other venture capital funds in new companies in growing markets—software, communications, Internet, and biotechnology.
Garage Technology Ventures	Venture capital investment bank providing private placement services for high-tech companies and investors in the communications, infrastructure, software, and wireless sectors, seeking to raise $2 to $15 million in a first or second institutional financing round.
Angstrom Partners	A joint venture between Lux Capital LLC, a venture firm focused on nanotechnology, and McGovern Capital LLC, an investment firm focused on technology development, capital formation, strategic transactions and alliances, and intellectual property value creation.

TABLE 4.2 (*Continued*)

Venture Capital Providers and Angels in the Nanotechnology Sector

Name of Organization	Services
Lux Capital	Venture capital firm of entrepreneurs and investors focused on identifying big breakthroughs, passionate entrepreneurs, transformative technologies, and massive markets, and investing, and in several cases participating as active board members, strategic advisers, or close extensions of entrepreneurial teams in emerging technology companies to solve global challenges in energy, technology, and healthcare.
Arrowhead Research Corporation	Sponsors nanoscience R&D in universities. In return for funding, Arrowhead obtains exclusive rights to license and commercialize technologies resulting from these R&D.
NGEN Partners	Catalytic, second-stage venture capital funding to emerging new businesses in the materials science and enabling technologies.
CW Group	Invests primarily in seed and early stage healthcare companies that it co-founds with leading scientists and entrepreneurs in biotechnology, pharmaceuticals, genomics, and diagnostics; medical devices and instrumentation; healthcare services and medical information systems.
Kleiner Perkins Caufield and Byers	Invests in early-stage and breakthrough ventures that have the promise to create new market opportunities.
Evolution Capital	Research driven investment bank that specializes in enabling technologies with expertise in diverse areas that include IP exploitation, emerging engineering, software and services, wireless and electronics, life sciences and nanotechnology to support private equity, corporate finance, equity capital markets.
Sevin Rosen Funds	Invests in early stage ventures with new technology products that can play market leadership roles.
Zone Ventures	Partnership funded by institutional investors for providing equity capital to young, high-growth companies.
California Technology Ventures	Helps start-up companies and investors by providing equity capital to early-stage technology companies, with special emphasis on Southern California companies.
Arcturus Capital	Early-stage venture capital firm specializing in development, commercialization, and financing of state-of-the-art technology companies.
Venrock Associates	Venture capital arm of the Rockefeller Family to fund entrepreneurs.
McGovern Capital	Originates, funds, structures, and implements capital formation, joint ventures, and business alliances.
Austin Ventures	Provides seed and early-stage investments in growth companies, primarily located in Texas and the Southwest.

(Continued)

TABLE 4.2 (*Continued*)

Venture Capital Providers and Angels in the Nanotechnology Sector

Name of Organization	Services
Quantiam Technologies	Develops and commercializes advanced materials based on nanotechnology; manufactures powders, catalysts, and coatings for the petrochemical, energy, and aerospace industries; provides consulting, technical, and research services for characterization of nanomaterials and surfaces; collaboration and seed investment in innovation.
NanoFrontier	Nanotechnology application development center and industrial resource center to work together with companies on joint projects to develop nanotechnology-enabled products, processes, and services.
Kim Ludvigsen	Specializes in financial engineering, in particular valuations, due diligences, and technology assessments.
JPMorgan Partners (JPMP)	A global partnership managing over $30 billion.
Hitachi CVC Fund	Invests at any stage of a company's development, including seed, first, and later rounds of financing through cash and in-kind investment by the Hitachi global network of corporations, alliance partners, venture capitalists, and customers.
NanoDimension	A Swiss-based nanotechnology investment firm.
Millennium Materials Technologies (MMT)	Funds for investments in the development and commercialization of novel advanced materials used in high growth industries like micro-electronics, pharmaceuticals, biotechnology, agro/chemicals, polymers, and communication materials.
FirstStage Capital	Corporate finance house that specializes in raising venture capital for early-stage technology companies.
Index Ventures	Works with high potential entrepreneurs to build technology companies with focused investments in European companies.
CenterPoint Ventures	An early-stage venture capital firm helping entrepreneurs to build successful, enduring companies.
Knowledge Market	Creates de novo start-ups and assists early-stage companies with early-stage funding, technology evaluation, technology transfer, strategy and positioning, and start-up operations including recruitment of key management, directors, and later-stage investors.
Advent Technologies	Commercialization of U.S. technologies into Japanese industry by leveraging relationships in the source side (U.S. R&D and technology organizations) and the target side (Japanese industry).
Israel Nanotechnology Trust	Fundraising and fund distribution arm of the Israeli National Nanotechnology Initiative (INNI)—the national initiative established by the government of Israel to set national priorities and goals for the advancement of nanotechnology in Israel.

TABLE 4.2 (*Continued*)

Venture Capital Providers and Angels in the Nanotechnology Sector

Name of Organization	Services
Pond Venture Partners	Invests in early-stage European companies in the semiconductor, communication, wireless, and software areas.
Qeam	Business consultancy and research company supporting organizations with innovation related to nanotechnology developments and commercialization.
NanoHoldings LLC NanoWorld Holding AG	Investment and commercialization company formed to assist University Tech Transfer Offices and their leading research scientists to commercialize key patentable inventions in the field of nanotechnology.
Headland Ventures LP	Venture partnership focused on early stage company investing.
Nanostart AG	Nanotechnology investment company.
Nanotech Partners	A private equity fund with sponsorship of Mitsubishi Corporation focuses exclusively on nanotechnology around the world.
Norwest Venture Partners	Offers its portfolio companies significant financial resources, seasoned business experience, and a network of relationships.
Mohr, Davidow Ventures (MDV)	Invests in innovative entrepreneurs with deep expertise in energy and materials, Internet services; life sciences, semiconductor software, and systems.
PureTech Ventures	Identifies scientific breakthroughs from research institutions, and provides management support and funding required to develop a venture or commercialization structure around the technologies that have been accepted into its portfolio.
Capital Stage Nanotech	Financial services provider with close links with its nanotechnology portfolio companies and an extensive network of contacts in industry and research institutions.
Intel Capital	Makes and manages financially attractive investments to create a portfolio of companies in support of Intel's strategic objectives, collaborates with the investment community at large, participates on boards in its portfolio companies, including offering entrepreneurs finance, its expertise, connections to Global 2000 companies and industry experts, programs, and overall brand value to create a distinct advantage for entrepreneurs.
Nanotech Capital	North Carolina-based nanotechnology intellectual property management and development company to assist inventors and entrepreneurs in the commercialization of patented and highly proprietary ideas.
3i	Global VC firm for nanotechnology focusing in private equity, infrastructure, and debt management.

(Continued)

TABLE 4.2 (*Continued*)

Venture Capital Providers and Angels in the Nanotechnology Sector

Name of Organization	Services
Global Crown Capital	Global Crown Capital, LLC, based in San Francisco, California, provides investment advice, brokerage, and asset management services to individuals, corporations, governments, endowments, and foundations. The company also offers wealth and portfolio management, hedge and mutual funds solutions, and risk management advisory services.
Seraphima Ventures	With expertise in assessment, development, and commercialization, it is focusing mainly on nanotechnology start-up companies with existing products and not in blue-sky research. Its business model is a hybrid of venture capital and angel investing.
VCgate	Directory of venture capital, private equity, merchant banking, and other investment firms from all over the world.
Innovest	Investment research and advisory firm specializing in analyzing companies' performance on environmental, social, and strategic governance issues, with a particular focus on their impact on competitiveness, profitability, and share price performance.

Source: Adapted from Nanotechnology Now, Nanotechnology Investment Companies: Venture Capital, Incubators, Angels, Wealth Management, Consulting Companies, and Recruitment and Placement Companies, http://www.nanotech-now.com/vc-firms.htm (July 28, 2011).

References

1. Youtie, J., Shapira, P., and Kay, L., Anticipating developments in nanotechnology commercialization, http://stip.gatech.edu/wp-content/uploads/2010/09/commercialization-nano-2010-8final.pdf.
2. Sylvester, D.J. and Bowman, D.M., Navigating the patent landscapes for nanotechnology: English gardens or tangled grounds? in *Biomedical Nanotechnology: Methods and Protocols*, Hurst, S.J.J., ed., Springer, New York, 2011, pp. 359–378.
3. Carlson, G. and DeKorte, D., The problem of patent thickets in convergent technologies, *Ann NY Acad Sci*, 1093, 180–200, 2006. Doi:10.1196/annals.1382.014.
4. Miller, J.C., Serrato, R., Represas-Cardenas, J.M, and Kundahl, G., *Handbook of Nanotechnology*, John Wiley & Sons, New York, 2004.
5. Duan, X., Huang, Y., Agarwal, R., and Lieber, C.M., Single-nanowire electrically driven lasers, *Nature*, 421, 241–24, 2003.
6. Serrato, R. and Chen, K., Mergers and acquisitions of nanotechnology companies: A review of the Unidym and CNI merger, *Nanotechnology Law & Business*, 4(2), 205–212, 2007.

5

Patent Litigations in Nanotechnologies

In the past two decades, nanotechnology has grown rapidly and is now poised for commercial takeoff in several fields leading to competitive posturing by the stakeholders. Being essentially technology-led, the key to leadership is strategic management of patent portfolios. It is therefore not surprising that battle lines have been drawn in the patent landscape to establish dominating positions through painstaking patent portfolio building, challenging validity of patents, enforcing patent rights, striking deals as an integral part of patent litigations, questioning contracts, and initiating proceedings based on competition law/antitrust laws.

There has been varying impact of litigations on nanotechnology business. Some businesses such as Evident Technologies, Inc., specializing in quantum dots and nanocrystals, filed a Chapter 11 bankruptcy protection and reorganization in the U.S. Bankruptcy Court of the Eastern District Court of New York on July 6, 2009. The major decision was catalyzed by a set of court decisions on patent and trademarks that went against Evident Technologies. It had a patent infringement case in U.S. District Court for Eastern District of Texas (*Invitrogen Corporation et al. v. Evident Technologies*). Evident Technologies was also involved in a trademark infringement suit (*Evident Technologies, Inc. v. Everstar Merchandise Company Ltd.*) in the U.S. District Court for Southern District of New York. In the trademark suit, the court dismissed with prejudice and with each party having to bear the cost of counsel. The cost of the cases and the lack of clients were the key reasons for Evident's monthly income of US$10,000 while the monthly expenses added up to US$200,000 a month.[1]

Through bankruptcy, Evident and Invitrogen reached a deal. Evident acknowledged it infringed Invitrogen's patent "in certain narrow limitations." Evident also agreed never to perform life sciences work. On March 30, 2010, a federal judge cleared Evident Technologies, Inc. to emerge from bankruptcy.[2] Nanotechnology is rich in such stories and this chapter exemplifies some of these trends through case studies.

The following cases are discussed.

- *Nanosys, Inc. v. Nanoco Technologies and Sigma-Aldrich Corporation*
- *DuPont Air Products NanoMaterials LLC v. Cabot Microelectronics Corporation*
- *Nano Proprietary, Inc. (NPI) v. Canon*

- *Elan Pharmaceuticals International Ltd. v. Abraxis Biosciences, Inc.*
- *Oxonica Energy Limited v. Neuftec Limited*
- *DSM v. 3D Systems*
- *Veeco Instruments v. Asylum Research*
- *Modumetal, Inc. v. Integran, Inc.*
- *MTS Systems Corp v. Hysitron, Inc.*
- *EV Group v. 3M*
- *Nanometrics, Inc. v. KLA-Tencor Corp*
- Other nanopatents litigation snippets

5.1 *Nanosys, Inc. v. Nanoco Technologies and Sigma-Aldrich Corporation*

Nanosys, Inc., founded in 2001, is a Delaware corporation having its principal place of business at 2625 Hanover Street, Palo Alto, California. Nanosys is an industry-leading nanotechnology company that has been developing, manufacturing, and selling products based on a technology platform of high performance inorganic nanostructures. Nanosys's technology is covered by a portfolio of over 670 patents and patent applications, including patents in the quantum dot field that are applicable applied in multiple industries including energy, electronics, optoelectronics, life science, and defense. Product areas of Nanosys's technology include flat-panel displays, nonvolatile memory, fuel cells, solid-state lighting, chemical analysis chips, and medical devices.

Nanosys had taken exclusive license from Massachusetts Institute of Technology (MIT) for U.S. Patent No. 6,861,155, issued March 1, 2005 entitled "Highly Luminescent Color Selective Nanocrsytalline Materials"; U.S. Patent No. 6,322,901, issued November 24, 2001 entitled "Highly Luminescent Color-Selective Nano-Crystalline Materials"; U.S. Patent No. 7,125,605, issued October 24, 2006 entitled "Highly Luminescent Color Selective Nanocrsytalline Materials"; U.S. Patent No. 6,821,337, issued November 23, 2004 "Preparation of Nanocrystallites"; and U.S. Patent No. 7,138,098, issued November 21, 2006 "Preparation of Nanocrystallites."

Nanoco Technologies Ltd. is a U.K. corporation, having its principal place of business at 46 Grafton Street, Manchester, United Kingdom.

Sigma is a Delaware corporation, having a principal place of business at 3050 Spruce Street, St. Louis, Missouri. Sigma is the U.S. and worldwide distributor for Nanoco's line of luminescent quantum dot nanocrystals, marketed under the Lumidot brand. As such, Sigma is the agent of its principal, Nanoco.

Nanosys filed a patent infringement suit in the U.S. District Court for the Western District of Wisconsin on April 27, 2009 against its U.K. rival Nanoco

Technologies Ltd., its U.S. distributor, Sigma-Aldrich Corp., over quantum dot technology claiming that the five Nanosys patents related to the technology were being infringed. Sigma-Aldrich was distributing Nanoco's quantum dot technology in the United States under the brand name Lumidots. Nanosys alleged that Nanoco's cadmium-based core shell (CdSe/ZnS) quantum dot technology distributed by Sigma-Aldrich was using quantum dot technology under patents licensed from MIT to produce devices such as solid-state lighting, photovoltaics, and electronic displays. Nanosys hoped for a final judgment in its favor holding infringement of the MIT quantum dot patents and issuance of an injunction barring Sigma-Aldrich from selling the dots in the United States. Further, Nanosys claimed that the defendants infringed willfully and therefore it was entitled to triple damages. The lawsuit claimed that the product being distributed in the United States infringes upon certain patents held by Nanosys.

The chief executive officer of Nanosys, Jason Hartlove, issued a statement:

> Nanosys is committed to serving visionary manufacturers by creating process-ready components for industries including electronics, energy efficiency, and medical devices. By enforcing ownership of our intellectually property, the manufacturers remain the real winners in having access to proven, trusted advanced material architecture, including quantum dot applications.

On July 23, 2009, Nanosys and Nanoco announced mutually agreed settlement terms of the patent infringement lawsuit. In the settlement, Nanoco agreed to terminate its current U.S. business for cadmium selenide core-shell quantum dots, but Nanoco did not admit that the asserted patents were infringed or valid. Additional terms of the settlement were not disclosed.

Following the settlement, Chief Executive Officer of Nanoco Michael Edelman made the following statement[3-5]:

> Nanoco is pleased with the outcome of the litigation. These products represent a tiny part of our business and we felt the small amount of product revenue affected by the litigation did not justify the expenditure of resources necessary to defend against a patent attack, Nanoco will continue to develop and deliver leading quantum dot products throughout the world with a specific focus on its heavy metal free technology.

5.2 *DuPont Air Products NanoMaterials LLC v. Cabot Microelectronics Corporation*

DA NanoMaterials LLC is a 50:50 joint venture between DuPont and Air Products. Headquartered in Tempe, Arizona with regional headquarters in

Hsinchu County, Taiwan, the company operates state-of-the-art applications and formulation laboratories in Tempe, Arizona, and Taiwan. It manufactures products in Asia, Europe, and North America to service the global semiconductor and wafer polishing markets. DA NanoMaterials markets its chemical mechanical polishing (CMP) slurries under the CoppeReady®, MicroPlanar®, Syton®, and Mazin™ trademarks. Additional information on the company can be obtained from its website (www.nanoslurry.com).

Cabot Microelectronics Corporation, headquartered in Aurora, Illinois, is the world's leading supplier of CMP slurries and is a growing CMP pad supplier to the semiconductor industry. The company's products play a critical role in the production of the most advanced semiconductor devices, enabling the manufacture of smaller, faster, and more complex devices by its customers. The company's mission is to create value by developing reliable and innovative solutions, through close customer collaboration, that solve today's challenges and help enable tomorrow's technology. Since becoming an independent public company in 2000, the company has grown to approximately 900 employees on a global basis

On December 6, 2006, DuPont Air Products NanoMaterials LLC filed a suit in the United States District Court for the District of Arizona, against Cabot Microelectronics, seeking a declaratory judgment that it did not infringe four U.S. Patents bearing Nos. 4,954,142 (the "142 patent"), 5,958,288 (the "288 patent"), 5,980,775 (the "775 patent"), and 6,068,787 (the "787 patent"), which are Cabot patents on CMP slurries. This suit followed an assertion of these patents by Cabot and extended negotiations between the parties.

DA NanoMaterials, a party to the suit, further stated in its suit:

- It has at all times relevant hereto manufactured and/or sold products of the type accused of infringement by Cabot and continues to manufacture and/or sell such products. Cabot's conduct has created on the part of DA NanoMaterials a reasonable apprehension that DA NanoMaterials will be faced with a patent infringement action if it continues to manufacture and/or sell its accused products.

- Cabot's conduct includes, *inter alia*, the sending of letters to DA NanoMaterials regarding DA NanoMaterials alleged infringement of the 142 and 288 patents.

- Cabot's conduct further includes, *inter alia*, allegations made by Cabot representatives to representatives of DA Nanomaterials, both during formal oral discussions and subsequent follow-up discussions, that DA NanoMaterials is allegedly infringing the 142, 288, 775, and 787 patents; contemporaneous demands by Cabot's representatives during these discussions that DA NanoMaterials cease-and-desist its allegedly infringing activities; and contemporaneous representations by Cabot's representatives during these discussions

that Cabot would not hesitate to bring an action for patent infringement if DA NanoMaterials did not cease-and-desist its allegedly
infringing activities.

- DA NanoMaterials has preliminarily explained to Cabot why it
  has not and does not infringe the 142, 288, 775, and 787 patents.
  Nevertheless, Cabot continues to assert that DA NanoMaterials has
  infringed and continues to infringe the 142, 288, 775, and 787 patents.

- DA NanoMaterials has, or will in the near future, inform Cabot that
  it will not cease-and-desist its allegedly infringing activities. Based
  on Cabot's actions and statements to date, DA NanoMaterials has
  a current, real, reasonable, and imminent apprehension that Cabot
  will immediately file a civil action for patent infringement against
  DA NanoMaterials upon receiving such information.

- As a result of Cabot's actions and statements, an actual justiciable
  controversy regarding the noninfringement, invalidity, and unenforceability of the 142, 288, 775, and 787 patents now exists.

Accordingly, DA NanoMaterials made a request to the court to enter judgment against Cabot for

- a declaration that DA NanoMaterials has not infringed and is not
  infringing the claims of the 142, 288, 775, and 787 patents;

- a declaration that each of the claims of the 288, 775, and 787 patents
  are invalid;

- a declaration that the 288, 775, and 787 patents are unenforceable;

- an injunction prohibiting Cabot from alleging infringement of the
  claims of the 142, 288, 775, and 787 patents by DA NanoMaterials;

- an award of damages DA NanoMaterials has sustained;

- a declaration that this case is an "exceptional case" within the meaning of 35 U.S.C. § 285 due to, *inter alia*, the above actions of Cabot;

- an award of costs and attorneys fees and other expenses DA
  NanoMaterials has been forced to incur; and

- such further relief as this Court may deem just and proper.

On January 2007, Cabot Microelectronics Corporation responded to the
declaratory-judgment complaint of DA NanoMaterials in three ways. First,
Cabot Microelectronics denied DA NanoMaterials' declaratory-judgment
claims, and responded to the complaint's specific allegations. Second,
Cabot Microelectronics asserted patent-infringement counterclaims against
DA NanoMaterials, in the counterclaims. Third, Cabot Microelectronics
asserted related patent-infringement claims against Precision Colloids, LLC
("Precision Colloids") in the third-party complaint.

Based on the explanations put forth by Cabot Microelectronics Corporation, it requested the court to

- Enter judgment in favor of Cabot Microelectronics on each of DA NanoMaterials' claims;
- Enter judgment in favor of Cabot Microelectronics on each of its counterclaims against DA NanoMaterials;
- Enter judgment in favor of Cabot Microelectronics on each of its claims against Precision Colloids;
- Adjudge that Cabot Microelectronics is the owner of the 288, 775, 787, 142, and 423 patents and all rights of recovery under each of them, and that each of these patents is good and valid in law and enforceable;
- Adjudge that DA NanoMaterials is infringing, has infringed, and has contributed to and has induced infringement of the 288, 775, 787, 142, and 423 patents, and that such infringement has been willful and deliberate;
- Preliminarily and permanently enjoin DA NanoMaterials, its affiliates, subsidiaries, officers, directors, employees, agents, licensees, successors, and assigns, and all persons in concert with them, from further infringement of the 288, 775, 787, 142, and 423 patents;
- Award Cabot Microelectronics compensatory damages from DA NanoMaterials, with interest;
- Award treble damages to Cabot Microelectronics for DA NanoMaterials' willful infringement;
- Require DA NanoMaterials to pay Cabot Microelectronics its costs and reasonable attorneys' fees under 35 U.S.C. § 285;
- Adjudge that Precision Colloids is infringing, has infringed, and has contributed to and has induced infringement of the 288, 775, 787, and 142 patents, and that such infringement has been willful and deliberate;
- Preliminarily and permanently enjoin Precision Colloids, its affiliates, subsidiaries, officers, directors, employees, agents, licensees, successors, and assignees, and all persons in concert with them, from further infringement of the 288, 775, 787, and 142 patents;
- Award Cabot Microelectronics compensatory damages from Precision Colloids, with interest;
- Award treble damages to Cabot Microelectronics for Precision Colloids' willful infringement;
- Require Precision Colloids to pay Cabot Microelectronics its costs and reasonable attorneys' fees under 35 U.S.C. § 285; and
- Award Cabot Microelectronics such other relief as the Court deems just and proper.

Cabot Microelectronics also demanded a jury for all claims in the complaint, counterclaims, and third-party complaint that are triable by jury. Cabot therefore alleged that DA NanoMaterials' manufacture and marketing of certain CMP slurries infringed on patents owned by Cabot. The affected DA NanoMaterials products include those used for tungsten CMP.

5.2.1 Court Proceedings

On July 25, 2008, the District Court issued its patent claim construction, or "Markman" Order ("Markman Order") in the litigation.[6] In a Markman ruling, a District Court hearing a patent infringement case interprets and rules on the scope and meaning of disputed patent claim language regarding the patents in the suit.

In the Markman Order, the District Court adopted interpretations of the terms used in the claims.

On January 27, 2009, Cabot Microelectronics filed a motion for summary judgment on DA NanoMaterials' infringement of certain of the patents at issue in the suit, and on that same date, DA NanoMaterials filed a motion for summary judgment on noninfringement and invalidity of certain of the patents at issue in the suit.

On November 16, 2009, the District Court issued its ruling on all of these respective summary judgment motions.[7]

In its summary judgment ruling, the District Court denied a motion filed by DA NanoMaterials for summary judgment of invalidity of three of the Cabot Microelectronics patents at issue in the case, which are fundamental patents in the field of tungsten CMP. The District Court also denied DA NanoMaterials' motion for summary judgment of noninfringement of these patents. In addition, the District Court denied DA NanoMaterials' motion for summary judgment of noninfringement of another one of Cabot Microelectronics patents at issue in the suit that is considered to be a foundational CMP patent. The District Court also denied Cabot Microelectronics' motion for summary judgment of infringement of the tungsten patents, stating that despite the weight of the record on DA NanoMaterials' infringement, summary judgment is not the forum to decide issues of fact that remain. The Court directed the parties to submit pretrial filings by December 16, 2009. The trial date was set for June 14, 2010.

On July 8, 2010, the validity of all of Cabot Microelectronics' patents was upheld in the company's ongoing patent enforcement litigation in the U.S. District Court for the District of Arizona against DuPont Air Products NanoMaterials, LLC ("DuPont Air Products NanoMaterials"). All of Cabot Microelectronics' patents at issue in the case were found valid. However, the jury found that DuPont Air Products NanoMaterials' products at issue did not infringe the asserted claims of these patents. The judgment order is provided in the appendices at the end of this chapter.

The next course of the proceedings and the strategy being followed by these two companies were being monitored with immense interest by the industry. Following these proceedings, in its financial statements for the third quarter of 2010, Cabot Microelectronics stated,

> Operating expenses in the third quarter of Fiscal 2010, which include research, development and technical, selling and marketing, and general and administrative expenses, were $34.5 million in the third fiscal quarter, or $9.5 million higher than the $25.1 million reported in the same quarter a year ago, driven primarily by higher variable incentive compensation costs, professional fees, including costs to enforce the company's intellectual property, other staffing related costs and travel expenses. Operating expenses were $2.4 million higher than the $32.1 million reported in the previous quarter, mostly due to higher costs to enforce the company's intellectual property. In total, the company spent approximately $3.7 million in the third fiscal quarter of 2010 to enforce its intellectual property against DuPont Air Products NanoMaterials, LLC and following the jury trial, which ended on July 8, the company expects its litigation costs to decrease significantly. Year to date, total operating expenses were $96.8 million. As stated in the company's July 13 announcement, the company now expects to exceed its previous guidance range for full year operating expenses of $120 million to $125 million for fiscal 2010.

"We are extremely pleased with the jury's conclusion that DA NanoMaterials' tungsten CMP slurries do not infringe Cabot's patents," said Seng Wui Lim, chief executive officer of DA NanoMaterials. "The ruling reinforces the fact that our products are vastly different from Cabot's and enables us to continue to aggressively pursue future business growth regarding these products. We remain committed to delivering innovative products, and will continue to compete in the marketplace worldwide while providing our customers with superior technology and service."

In addition to this court ruling, on June 18, the appellate Korean Patent Court found Cabot's tungsten CMP Korean patent invalid. This continues a series of losses by Cabot in Korean court cases finding Cabot's tungsten CMP patent invalid, including invalidations in the Seoul District Court, the appellate Korean High Court, and the Korean Intellectual Property Tribunal.[8–11]

5.3 *Nano Proprietary, Inc. (NPI) v. Canon*

Applied Nanotech (AN), a subsidiary of Nano Proprietary, Inc. (NPI) of Austin, Texas, ran an IP (royalty)-based business model. It acquired intellectual property by way of patents, trademarks, know-how, etc. and, after further developing the patented technologies, it strategically licensed the improved technology under the patents for royalties from the licensee based on sale of products by the licensee.

Canon, a Japanese corporation, produces, *inter alia*, copiers, printers, and cameras. In 1997, Canon began discussions with Toshiba, another Japanese corporation and leading television manufacturer, about joint development of field electron emission display (FED) televisions.

The patents in question are core nanotube patents describing CNT FED televisions and surface electron (slit stimulated) emission TV displays (SED). CNTs emit electrons similar to an electron gun in a TV tube firing electrons at screen phosphors. Printing CNTs and control circuitry just behind each phosphor results in very thin displays, low manufacturing cost, and incredible sharpness. SEDs use a nanosized slit with or without nanotubes for a similar effect. AN claimed it has patents in both areas.

5.3.1 The Issue

Canon licensed the SED patent technology (lump sum limited rights purchase) from AN and formed a joint venture (SED, Inc.) with Toshiba, one of the largest TV display makers to create and market SED TVs. As a part of the joint venture, Canon shared CNT/SED information with Toshiba. AN, claiming that Canon had no right to share its patented information with a nonsubsidiary, sued to enjoin any product made by Toshiba or by the joint venture. The District Court found that Canon materially breached the patent license agreement via an impermissible sublicense to SED, such that AN was entitled to terminate the agreement. The issue of damages was tried before a jury, which found that AN was not entitled to any damages based on the value of a prospective license. Because of the suit and uncertainties of the outcome of the litigation, Toshiba indefinitely postponed product introduction of its new 55-inch SED screen TV.

Following is the chronological progression of the matter in the courts:

- In December 1998, Nano and Canon began negotiations regarding licensing Nano's FED patents. During these negotiations, Canon did not mention its ongoing discussions with Toshiba. On March 26, 1999, Nano and Canon executed a patent license agreement (PLA) that granted Canon and its subsidiaries a nonexclusive license to Nano's FED patents. Canon paid a one-time lump sum of $5,555,555.55 and received a "fully paid-up, worldwide, royalty-free, irrevocable, perpetual, nonexclusive license (without the right to sublicense)" that "shall continue in full force and effect until expiration of the last to expire of the licensed patents." The PLA further provided that it "shall be construed by and interpreted in accordance with the laws of the state of New York, United States of America, exclusive of its choice of law provisions."

- On June 9, 1999, Canon and Toshiba executed a written Joint Development Agreement and began joint research and development work in Japan with the companies having equal control.
- In 2004, Canon and Toshiba began discussions about forming a joint venture and on September 14, 2004, Canon and Toshiba executed a Joint Venture Agreement (JVA) providing that, at all times, Canon's shares in SED must exceed Toshiba's shares by one share. It was also recorded that SED was involved in R&D related to FED patents in Japan but had not produced or marketed any products.
- On April 11, 2005, AN filed suit against Canon and Canon USA in the U.S. District Court for the Western District of Texas asserting: (1) Canon materially breached the PLA by sublicensing its patents to SED/Toshiba; and (2) Canon tortiously interfered with Nano's prospective business relations (specifically, its prospective license with SED/Toshiba). In addition, Nano sought a declaratory judgment that SED did not qualify as a subsidiary under the PLA.
- In October 2005, the district court dismissed AN's count for tortious interference with prospective business relations pursuant to Federal Rule of Civil Procedure 12(b)(6).
- In April 2006, AN amended its complaint, adding claims that Canon committed fraud during the PLA negotiations and seeking rescission of the PLA. Canon moved for partial summary judgment on AN's breach of contract and declaratory judgment claims on the grounds that SED qualified as a Canon subsidiary, which the district court denied. The district court found that SED was not a subsidiary of Canon because Canon did not hold a majority of "stock conferring the right to vote at general meetings" and, alternatively, because the court declined on equitable grounds "to recognize a corporate fiction designed for the sole purpose of evading Canon's contractual obligations."[12] Specifically, the district court found that Canon's agreement "not to use its majority share to outvote Toshiba on matters governed by the JVA" meant that Canon "does not hold a majority of 'stock conferring the right to vote at general meetings.'"
- On December 1, 2006, AN informed Canon that it was terminating the PLA.
- Canon disputed AN's right to terminate and informed AN that its purported termination was ineffective. Subsequently on January 12, 2007, Canon and Toshiba executed a Stock Transfer Agreement, and on January 29, Canon bought all of Toshiba's stock in SED for approximately $83 million—the same amount that Toshiba had originally paid thereby making SED a 100%-owned subsidiary of Canon.
- Canon then moved for summary judgment that SED, in its new structure, was a Canon subsidiary. AN moved for summary judgment

regarding whether Canon had breached the PLA and whether its termination of the PLA was effective.

- On February 22, 2007, the U.S. District Court for the Western District of Texas granted AN's motion, finding that: (1) SED in its original form was not a Canon subsidiary; (2) Canon had materially breached the PLA because creating SED "was effectively an attempt to sublicense its rights to the AN patents"; (3) AN was damaged by this breach (although the district court did not assign a value to this damage); (4) AN's termination of the PLA was effective; and (5) Canon's restructuring of SED was ineffective to prevent termination because it "was not undertaken within a reasonable time."[13] The district court permitted AN to retain the approximately $5.5 million lump sum payment from Canon. Furthermore, the district court noted that it "did not see how there were provable consequential damages," but nonetheless permitted AN to attempt to prove such damages at trial.[14]

- From April 30 to May 3, 2007, AN's remaining fraud and damages claims were tried before a jury. At the close of AN's evidence, Canon moved for judgment as a matter of law on the fraud claim, and AN voluntarily dismissed its fraud claim. The damages claim went to the jury, and it returned a verdict that AN had not sustained any "damages from Canon's conduct in addition to the termination of the patent license agreement and retention of the $5.5 million purchase price." Thus, AN took nothing further as a result of the jury trial. Canon timely appealed the district court's summary judgment orders, and AN timely cross-appealed various trial rulings and the dismissal of its tortious interference with prospective business relations claim.

- On July 25, 2008, the United States Court of Appeals, Fifth Circuit on detailed consideration of the matter observed:
  - AN was not entitled to terminate the license because it contracted for an irrevocable and perpetual license. At the time of material breach, AN was entitled to sue Canon under any available form of relief, but termination of the PLA was not permissible.
  - SED qualifies as a subsidiary under the definition provided in Section 1.3 of the PLA. As a result, SED's use of the FED patents is permissible under its current ownership structure.
  - AN's arguments lack merit and affirm the district court's judgment regarding the cross-appeal in all respects and the district court's ruling was not an abuse of discretion.
  - The district court properly precluded testimony about a correlation between fluctuations in AN's stock and any lost license as too speculative.

- AN is not entitled to damages based on the value of a prospective license because it did not prove such damages "with reasonable certainty." Therefore, even if AN is correct that the jury instructions were erroneous, it is not entitled to a remand on this basis.

- AN is correct that—under one line of cases—it need not identify a specific contract that would have occurred. However, the "business relationship" versus "contractual relationship" distinction is of no consequence here because AN presented no evidence of a reasonable probability of either an impending business relationship or a contractual relationship. Therefore, the district court's dismissal of AN's tortious interference with prospective business relations claim is affirmed.

The U.S. Court of Appeals for the Fifth Circuit issued its opinion in which it in part affirmed, and in part reversed, the rulings of the district court. While the appeals court accepted, without deciding, the district court's decision that SED, Inc. as originally formed did not qualify as a Canon subsidiary, and that Canon had materially breached the contract, it found that termination of the license agreement was not an appropriate remedy.

The appeals court also ruled that the restructured SED, Inc., which is 100% owned by Canon, now qualifies as a Canon subsidiary. The appeals court denied AN's appeal that the district court had improperly excluded certain evidence from the trial. The decision reinstates Canon's nonexclusive license to substantially all of AN's field emission patents, excluding certain display applications.

AN did not appeal the court's decision that was ruled against them and cleared the way for the production of SED TV by Canon. On December 2, 2008, AN dropped the lawsuit, stating that continuing the lawsuit "would probably be a futile effort."

Although SED failed to show at three consecutive Consumer Electronics Shows (CESs), on April 30, 2009 Canon filed two patents for SED; namely, U.S. Patent Application 20090108727 and U.S. Patent Application 20090111350. However, in early 2011 Canon halted any further development of SED displays for home use stating its inability to bring down production costs as the main reason for stopping further SED TV work.[15–20]

5.4 *Elan Pharmaceuticals International Ltd. v. Abraxis BioScience, Inc.*

Abraxis BioScience, headquartered in Los Angeles, California, is a fully integrated biotechnology company dedicated to delivering progressive therapeutics, such as ABRAXANE, and core technologies that offer patients and medical professionals safer and more effective treatments for cancer and other critical illnesses. Abraxis' portfolio includes the world's first and only protein-based nanoparticle chemotherapeutic compound. Scientists at Abraxis developed its proprietary tumor targeting system known as the nab™ technology platform, an innovative approach to treating cancer and other critical illnesses. By harnessing the unique natural properties of the human protein albumin, Abraxis' nab technology platform enables the transport and delivery of therapeutic agents to the site of disease. This revolutionary biologically interactive delivery system is being further developed with other water-insoluble drugs for their potential use across a broad range of tumors. Abraxane® for Injectable Suspension (paclitaxel protein-bound particles for injectable suspension) (albumin bound), the first U.S. Food and Drug Administration (FDA) approved product to use the nab platform, was launched in 2005 for the treatment of metastatic breast cancer.[21]

Abraxis BioScience, Inc. and its subsidiaries became a wholly owned subsidiary of Celgene Corporation in July 2010 when it was bought for $2.9 billion cash and stock.

Elan is a neuroscience-based biotechnology company headquartered in Dublin, Ireland. It was incorporated as a private limited company in December 1969 and became a public limited company in January 1984. Elan's principal research and development, manufacturing, and marketing facilities are located in Ireland and the United States. Elan's bioneurology work includes research, development, and commercial activities for neurodegenerative diseases, such as Alzheimer's disease, Parkinson's disease, and autoimmune diseases, including multiple sclerosis.

Elan is the owner of two patents, U.S. Patent Nos. 5399363 (363) and 5,834,025 (025).

The 363 patent essentially claims the following:

- Particles consisting essentially of 99.9% by weight of a crystalline medicament useful in treating cancer susceptible to treatment with said medicament, said medicament having a solubility in water of less than 10 mg/ml, and having a noncrosslinked surface modifier adsorbed on the surface thereof in an amount of 0.1 to 90% by weight and sufficient to maintain an average effective particle size of less than 1000 nm, wherein said medicament is selected from the group consisting of and adrenocortical suppressants
- A composition comprising the said particles
- A method of administering an effective amount of anticancer composition comprising the said particles
- A method of treating a mammal comprising administering to the mammal an effective amount of an anticancer agent, the improvement wherein the efficacy of said anticancer agent is increased by administering said anticancer agent in the form of the said particles

The 025 patent essentially claims the following:

- Methods of intravenous administration of nanoparticulate drug formulations to a mammal to avoid adverse hemodynamic effects by
  reducing the rate and concentration of the nanoparticles in the formulations, by pretreating the subject with histamine, or by pretreating the subject with a desensitizing amount of the nanoparticulate
  drug formulations

In July 2006, Elan filed a suit against Abraxis in the United States District
Court of Delaware (No. 1:06-cv-00438) claiming that the Abraxane reformu-
lation of the traditional breast cancer treatment paclitaxel (Taxol), marketed
in the United States since 2005 by Abraxis, directly infringes two of its patents by Abraxis making and selling Abraxane. In its suit, Elan alleged that
Abraxane (protein nanoparticles) enhances the delivery of poorly water-
soluble compounds such as taxol.

Elan also claimed that Abraxis is liable for contributory and inducing
infringement and that Abraxis is liable for the direct infringement by medical professionals who administer Abraxane to patients.

Elan requested the court for injunctive relief, treble damages for willful
infringement, and attorneys' fees.

In its response, Abraxis denied infringement of Elan's patents and further
challenged the validity the two patents 363 and 025 on the grounds of obviousness in light of the prior art.

Abraxis alleged that the patents are invalid because they were rendered
obvious by prior art and that they were unenforceable as the applicant had
deliberately not cited a critical reference to the USPTO.

Both parties made detailed submissions by way of evidences. During the
discovery stage, many discussions took place on the word "crystalline" and
to ascertain whether an amorphous paclitaxel composition infringes under
the doctrine of equivalents.

Further, during the discovery stage, several issues were raised with regard
to a demand by Elan for disclosure by Abraxis of its manufacturing process.
Abraxis objected to such disclosure of its manufacturing process.

However, the court by a first interim order dated June 12, 2007 permitted
specified Elan outside counsel expedited access to secret Abraxane manufacturing information while the parties negotiated an Amended Stipulated
Protective Order.

Subsequently via a second interim order in July 2007, the court directed that
Dr. Robert O. Williams and one additional Elan expert witness ("Approved
Elan Experts") be provided access to secret Abraxane manufacturing information for private electronic viewing through the third-party Intralinks
website at any of the U.S. offices of Baker Botts LLP in the presence of any of
five identified outside attorneys for Elan (specifically, Stephen Scheve, Paul F.

Fehlner, William J. Sipio, Robert Riddle, and Jennifer Cozeolino) ("Approved Elan Outside Counsel"). Each approved Elan expert was directed to execute the written agreement in Exhibit A annexed to the judgment and provide the original signed agreement to counsel for Abraxis before having access to any of the secret Abraxane manufacturing information.[22]

This was followed by a debate on the claims construction. The judge issued his claims construction order in which the Court concluded that a largely amorphous product could not infringe the claim term "crystalline medicament."

An elaborate account on the various submissions and the techno-legal complexities including the pretrial order and jury instructions are available in Endnotes 23 through 27.

The questions before the jury on June 2, 2008, on matters related to Infringement, Willful Infringement, Validity, Damages, and Unenforceability (inequitable conduct, unclean hands) were as follows:

I. As to 363 Patent Infringement

- Has Elan shown by preponderance of the evidence that the particles in Abraxane consist essentially of 99.9 to 100% by weight of a crystalline medicament and 0.1 to 90% by weight of a surface modifier that is essentially free of intermolecular cross linkages?
- Has Elan shown by preponderance of the evidence that the particles in Abraxane consist essentially of 99.9 to 100% by weight of a crystalline medicament under the doctrine of equivalence and 0.1 to 90% by weight of a surface modifier that is essentially free of intermolecular cross linkages?
- If you answered "yes" as to either question 1 or 2 based on claims 3, 5, 10, or 11, has Elan shown by preponderance of evidence that Abraxis induced infringement of claims 3, 5, 10, or 11, by others?
- If you answered "yes" as to either question 1 or 2 based on claims 3, 5, 10, or 11, has Elan shown by preponderance of evidence that Abraxis contributorily infringed claims 3, 5, 10, or 11, by others?

Willful Infringement

- If you answered "yes" to any of the questions 1 through 4 for claims 3, 5, 10, or 11, do you find that Elan has shown clear and convincing evidence that Abraxis wilfully infringed the 363 patent?

Validity

- Has Abraxis shown by clear and convincing evidence that any of the following claims 1, 3, 5, 10, 11 of the 363 patent are invalid due to obviousness?

- Has Abraxis shown by clear and convincing evidence that any of the following claims 1, 3, 5, 10, 11 of the 363 patent are invalid for failure to satisfy enablement requirement?
- Has Abraxis shown by clear and convincing evidence that any of the following claims 1, 3, 5, 10, 11 of the 363 patent are invalid for failure to satisfy the written description requirement?
- Has Abraxis shown by clear and convincing evidence that any of the following claims 1, 3, 5, 10, 11 of the 363 patent are invalid for failure to satisfy the definiteness requirement?
- Has Abraxis shown by clear and convincing evidence that any of the following claims 1, 3, 5, 10, 11 of the 363 patent are invalid due to obviousness-type double patenting?

Damages

If you have found at least one of the claims 3, 5, 10, or 11 of the 262 patent is infringed and valid, what amount of damages in the form of a reasonable royalty do you find Elan has proven by preponderance of evidence that it is entitled to received from Abraxis?

If from October 26, 2007 Amount $
If from February 2005 Amount $

Inequitable Conduct

Has Abraxis shown by clear and convincing evidence that the 363 patent is unenforceable due to inequitable conduct?

Unclean Hands

Has Abraxis shown by clear and convincing evidence that the 363 patent is unenforceable due to unclean hands?

II. As to 025 Patent

Validity

- Has Abraxis shown by clear and convincing evidence that any of the following claims 1, 2, 3, 13, 14, 15 of the 025 patent are invalid for failure to satisfy enablement requirement?
- Has Abraxis shown by clear and convincing evidence that any of the following claims 1, 2, 3, 13, 14, 15 of the 025 patent are invalid for failure to satisfy the written description requirement?
- Has Abraxis shown by clear and convincing evidence that any of the following claims 1, 2, 3, 13, 14, 15 of the 025 patent are invalid for failure to satisfy the definiteness requirement?

Inequitable Conduct

- Has Abraxis shown by clear and convincing evidence that the 025 patent is unenforceable due to inequitable conduct?

5.4.1 Jury Verdict

On June 13, 2008, after 10 days of trial, the jury delivered a verdict finding the 363 patent infringed and awarded Elan damages of $55.2 million based on a reasonable royalty rate of 6%. The jury also rejected defenses based on lack of enablement, lack of adequate written description, and inequitable conduct.[28]

5.4.2 The Judgment[29]

This action came before the Court for a trial by jury. The issues have been tried and the jury rendered its verdict on June 13, 2008. The verdict was accompanied by a verdict form (D.I. 613), a copy of which is attached hereto. Therefore, IT IS HEREBY ORDERED AND ADJUDGED that judgment be and is hereby entered in favor of the plaintiff, ELAN PHARMA INTERNATIONAL LTD and against the defendant, ABRAXIS BIOSCIENCE, INC., that ABRAXIS BIOSCIENCE, INC., literally infringes Claims 3 and 5 of U.S. Patent No. 5,399,363 in the amount of FIFTY-FIVE MILLION TWO HUNDRED THIRTY THOUSAND DOLLARS ($55,230,000.00) for damages from January 7, 2005 through June 13, 2008, FINDING A REASONABLE ROYALTY RATE OF SIX PERCENT (6%); and that the infringement is not willful; IT IS FURTHER ORDERED that judgment be and is hereby entered in favor of the plaintiff, ELAN PHARMA INTERNATIONAL LTD and against the defendant, ABRAXIS BIOSCIENCE, INC., for the validity of Claims 1, 3, 5, 10, and 11 of U.S. Patent No. 5,399,363 and Claims 1, 2, 3, 13, 14, and 15 of U.S. Patent No. 5,834,025;

IT IS FURTHER ORDERED that judgment be and is hereby entered in favor of the plaintiff, ELAN PHARMA INTERNATIONAL LTD and against the defendant,

ABRAXIS BIOSCIENCE, INC., for the enforceability of U.S. Patent Nos. 5,399,363 and 5,834,025.

Dated: June 16, 2008

/s/ Gregory M. Sleet UNITED STATES DISTRICT CHIEF JUDGE

5.4.3 The Final Settlement

On February 24, 2011, Elan Corp Plc and Celgene Corp said they settled a patent infringement lawsuit over Celgene breast cancer drug Abraxane. Under the settlement, Celgene will pay Elan a one-time fee of $78 million and will receive the worldwide license to several Elan U.S. and foreign patents for the drug. Elan will not receive any future payments for the drug.

Abraxis, based in Los Angeles, stock saw its stock fall $1.79, or 2.7%, to $65.61 with the breaking of the news on February 25, 2011 at 4:30 p.m. EST in Nasdaq Stock Market composite trading. The stock had fallen 4.6% in 2011. Elan stock which was up 14% in 2011, by the time of the settlement fell 28 cents (euro) to 16.82 euros in Dublin.

5.5 *Oxonica Energy Ltd. v. Neuftec Ltd.*

The matter relates to a dispute on an exclusive patent and know-how license between Oxonica and Neuftec relating to fuel additives.

Oxonica is an advanced materials business with its headquarters in Haddenham, U.K. It was a spin-off from Oxford University, England in 1999 and has been built on a foundation of solid scientific procedures and protocols. It has developed a range of technology-led businesses based on nanomaterials. It has restructured its businesses from time to time by selling off its subsidiaries and striking strategic alliances with other technology-led companies.

Neuftec is a research and development company that owns patents in the fuel catalyst technology sector used in the production of the first version of Envirox (2001–2007) that was manufactured by Advanced Nanotechnology of Australia under a manufacturing license granted by Oxonica with authorization from Neuftec. On February 29, 2008, IP Australia issued a re-examination report with respect to Neuftec's Australian patent. The patent had been issued in 2006 with 10 claims. In the re-examination report, IP Australia confirmed the patentability of the 10 claims as originally granted in 2006 in re-examination proceedings initiated by Oxonica. Neuftec was also opposing the grant of an Oxonica patent relating to doped cerium oxide fuel additives in Australia. The Neuftec technology was then being utilized by Energenics Pte Ltd. in its Enercat fuel additive.

In 1996, Bryan Morgan started a company called Celox Ltd., which was the predecessor in title of Neuftec. Celox was interested in fuel catalysts. In December 1999, Ronendra Hazarika became a director of Celox. Hazarika had a background in the motor racing industry. Celox decided to produce an additive for diesel fuel—a catalyst that could be dispersed in the fuel and would reduce fuel consumption and engine emissions. Celox started to do research using lanthanide oxides and, in particular, cerium oxide as fuel additives.

However, the difficulty with earlier lanthanide-based additives was that they were known to clog the filters and cause abrasion and blockages in engines. Celox thought of coating the cerium oxide particles with lipophilic materials to make them disperse in the fuel.

Celox approached Oxonica, which had knowledge of nanomaterials, and the two companies exchanged information and knowledge under a confidentiality agreement. A joint venture was proposed wherein Celox would provide its knowledge about fuel additives and the markets for such additives and Oxonica would provide its knowledge about nanoparticles. A joint venture company was incorporated. Celox Ltd. and Oxonica were to be 50:50 owners and to share the benefits accordingly. Oxonica initiated research on such fuel additives.

Celox Ltd. had applied for patent protection in the United Kingdom on June 29 and September 13, 2000. Those patent applications were withdrawn and were not published, but they served as a basis to claim international priority. On June 29, 2001, Celox made a patent application under the Patent Cooperation Treaty (PCT) filed as Application No. PCT/ GB01/02911 on June 29, 2001 claiming priority of the U.K. patent filings GB 0016032.5 dated June 29, 2000 and GB 0022449.3 dated September 13, 2000. The PCT application described and claimed the substances that could be used as fuel additives and with further disclosures and claims that these fuel additives should be particles of lanthanide oxides of 1 to 50 nm coated with lipophilic agents. The lipophilic agents were also disclosed and claimed.

The PCT application was assigned by Celox to Neuftec. Morgan and Hazarika were also directors of Neuftec Ltd.

On December 7, 2001, an arrangement between Neuftec and a "subsidiary of Oxonica" named "RMBKNE5 Limited" was affected by two documents: (1) a Main Agreement covering the operation of the business and (2) an exclusive license (the License Deed). The Oxonica subsidiary was made a party to the Main Agreement and was the licensee under the License Deed.

The PCT was published on January 3, 2002. At some date in 2002 with the consent of Neuftec, the prosecution of patent applications was taken over by Oxonica and paid for by it, but it kept Neuftec informed.

National phase filings were done based on the PCT application and patents were granted to Neuftec in several countries. Applications in some countries were pending. However, in several countries the claims granted were substantially narrower than those in the PCT application as the examiners found relevant prior art. For example, the broadest claim in the granted European Patent 1,299,508 granted on January 12, 2005 provides for a method of improving the efficiency with which fuel is burnt or of reducing emissions, said method comprising dispersing an amount of at least one particulate lanthanide oxide in the fuel, wherein the lanthanide oxide is coated with an alkyl carboxylic anhydride.

This should be contrasted with Claim 1 in the PCT application, which read as

> A method of improving the efficiency with which fuel is burnt in a fuel burning apparatus and/or a method of reducing the emissions produced by a fuel which is burnt in a fuel burning apparatus, said method comprising dispersing an amount of at least one particulate lanthanide oxide in the fuel.

It should be noted that after the examination the claim in the granted European Patent has been narrowed to lanthanide oxide coated with an alkyl carboxylic anhydride.

Similarly in Australia and other countries, the claims have been narrowed during the examination of the PCT application in light of the prior art. It is relevant at this stage to read the main provisions of the License Deed signed on December 7, 2001. Two aspects are of significance:

1. Neuftec has developed a fuel additive invention for which an international patent application (number PCT/GB01/02911) was filed at the U.K. Patent Office on June 29, 2001, a copy of which is attached in the Schedule hereto.
2. Neuftec, the Licensee (i.e., Oxonica Subsidiary) and Oxonica Ltd. (i.e., Oxonica Parent) have entered into an agreement dated December 7, 2001 (the "Main Agreement") under which Neuftec agrees to grant and the Licensee agrees to accept an exclusive license under the patent application to manufacture, use, and sell certain products, as defined below, on the terms set out in this Deed.

The operative part of the License Deed dealing with the definitions and conditions for the grant of the license are also of significance.

Definitions

"Licensed Application"	The application appended in the Schedule hereto (i.e., the PCT) and any continuation, continuation-in-part, or divisional applications thereof as well as foreign counterparts and re-issues thereof.
"Licensed Know-How"	All technical information owned or possessed by Neuftec at the date of this Deed and thereafter, whether patentable or otherwise, relating to the combustion of fuels using cerium oxide, which information is necessary or useful for the development, manufacture, use, and/or sale of the Licensed Products hereunder.
"Licensed Patent"	Any patent issuing from the Licensed Application thereof as well as foreign counterparts and re-issues thereof.
"Licensed Products"	Any product, process, or use falling within the scope of claims in the Licensed Application or Licensed Patent.
"Net Sale Price"	The price paid by a purchaser of the Licensed Products excluding all discounts and costs attributable to freight, distribution, and import duty.
"Option"	An option in favor of Neuftec, whereby Neuftec may terminate this Deed if Neuftec has not received from the Licensee the sum of £250,000 due and payable under Clause 7.2 of the Main Agreement.
"Territory"	The world.

Grant of License

- In consideration of the Licensee (i.e., Oxonica Subsidiary) agreeing to pay to Neuftec the payments set out in Clause 4, Neuftec hereby grants the Licensee an exclusive license under the Licensed Know-how, the Licensed Application, and/or Licensed Patent in all fields to make, manufacture, use, sell, and/or otherwise exploit the Licensed Products throughout the Territory.

- This Deed shall continue in force until either the date on which (a) the last of the Licensed Patents has expired or (b) the last of the Licensed Applications has been revoked or (c) the Licensee goes into liquidation or (d) the Main Agreement is terminated.
- Notwithstanding Clause 2.2, this Deed shall terminate with immediate effect upon Neuftec serving written notice on all Parties hereto that Neuftec is exercising the Option.
- Upon the termination of this Deed, the Licensee shall return all Licensed Know-How and shall cease to exploit the Licensed Products. (This does not apply to developments, modifications, or improvements made by the Licensee.)

Royalty Fees

- During the Term the Licensee shall pay to Neuftec a royalty based on 5% of the Net Sale Price received by the Licensee (the "Neuftec Royalty Fees").
- The Licensee must keep accurate records showing the information used by it to calculate the royalties and these may be inspected.
- The Neuftec Royalty Fees shall be payable every six months upon the revenue received by the Licensee from the sale of the Licensed Products.
- If the Annual Accounts show that the turnover attributable to the Business is £10 million or more Oxonica shall pay to Neuftec the sum of £1 million (the "Revenue Milestone Payment").
- If the Annual Accounts show that there is a positive Cumulative Profit, Oxonica shall pay to Neuftec 25% of the amount of the Annual Profit shown in the Annual Accounts.
- The payments referred to in Clauses 4.5 and 4.6 shall be due and payable provided the Annual Accounts show a positive cash flow.

Hon. Peter Prescott QC (sitting as a Deputy Judge) in his judgment[30] summarized the main benefits for the Oxonica and Neuftec parties, respectively, as:

Benefits for Oxonica

- An introduction to the general business opportunity.
- Neuftec's know-how and the exclusive right to exploit it.
- The exclusive right to exploit the Neuftec patent rights.
- No competition from Neuftec (irrespective of the breadth of the patent rights).

Benefits for Neuftec

- A 5% royalty on sales of licensed products.
- A lump sum of £250,000 if the Investment Round is completed within 6 months.
- If and when there is a positive cash flow, 25% of Oxonica's annual profits, provided there is a positive cumulative profit.
- A lump sum of £1 million if and when the business grows to a certain size.
- The exclusive right to exploit the Neuftec patent rights.
- No competition from Neuftec (irrespective of the breadth of the patent rights).

5.5.1 The Issue

In June 2006, Oxonica was in negotiation for a contract with a Turkish oil company for a supply of fuel additives "Envirox," which would involve payments to Neuftec under the License Deed. On October 5, 2006, Oxonica informed Neuftec that Oxonica had decided to obtain the Envirox to be supplied under the Turkish contract from a different source. Further, Oxonica claimed new fuel additives "Envirox 2," which were outside the ambit of claims of Neuftec's granted patents as the coating of particles of rare earth oxide not belong to the delimited class, that is, alkyl carboxylic anhydrides (even though Envirox 2 would fall within the ambit of the claims of the PCT application) and, therefore, Neuftec would not be entitled to any payments from Oxonica.

Oxonica filed a suit in the High Court of Justice Chancery Division Patent Court U.K. for a declaration that Envirox 2 is not a Licensed Product as defined and that sales of it in the United Kingdom or in any other territory to which a granted patent relates with claims in the same or substantially the same form as European Patent (U.K.) 1,299,508 do not attract royalties under the Licensed Deed. Neuftec counterclaims for an audit in respect of sales of Envirox 2 and payment of all sums due.

The key issues before the court were

- The meaning of the expression Licensed Products, upon the sale of which royalties are payable in the context of the licensing agreement; that is, whether "any product, process or use falling within the scope of claims in the Licensed Application or Licensed Patent" would qualify as the licensed product.
- What would be the role of "know-how" in the license agreement?

5.5.2 The Judgment

Hon. Justice Peter Prescott QC considered all the submissions made by the two parties and deliberated on

- The definition of "licensed products" as meaning "any product, process or use falling within the scope of claims in the Licensed Application or Licensed Patent" (two alternatives).
- The onset of the know-how even if the patent applications are granted or not granted and under what circumstances the product Envirox 2 would fall within the definition of "the licensed product" and payment of royalties based on sales of licensed products.

Based on a detailed analysis, Hon. Justice Peter Prescott in his judgment dated September 5, 2008 stated:

> Royalties are payable in respect of any product, process or use falling within the scope of any claim of the PCT application as appended to the License Deed, and nothing else. Envirox 2 is a Licensed Product as defined, and attracts royalties accordingly. The claim fails and the counterclaim succeeds.
>
> If my interpretation of the License Deed is wrong, it still would not be appropriate to grant the declaratory relief sought by Oxonica. The answer might then depend on the existence and status of divisional applications and the like.
>
> I shall hear the parties about the form of relief or they can, if they wish, send their submissions by email to my clerk by some convenient date to be arranged. I am presently minded to say that (1) costs should follow the event and (2) permission to appeal, if sought, is almost inevitable.

Oxonica appealed against this judgment before the Supreme Court of Judicature, Court of Appeal (Civil Division). The matter was heard before the President of the Queen's Bench Division, the Rt Hon. Lady Justice Arden and the Rt Hon. Lord Justice Jacob.[31]

The question for the Court of Appeal was the meaning of the phrase "claims in the Licensed Application or Licensed Patent." Did it mean (1) any product covered by the claim of the PCT application (currently the widest claim), (2) any product covered by the claims of either the PCT application or any later patent, even where the later patent's claim might be wider, or (3) any product covered by the claims of either the PCT application or a national application as the case might be?[32]

The Hon. Lady Justice Arden and Hon. Justice Jacob upheld the High Court's decision in favor of meaning (1) and further concluded that the licensing agreement was poorly drafted and dismissed the appeal and stated that Hon. Justice Peter Prescott in the first instance had been right to reject Oxonica's construction. Further, they said it did not make sense for Oxonica to have free use of Neuftec's know-how in countries where there was a more restricted patent. Although Oxonica might have to compete against third parties in such countries, the third parties would have to develop their own know-how, and the parties clearly considered the know-how to have value. Furthermore, it was unlikely that the parties had intended the patent situation to be determined for each sale, as this "would be in a constant state of flux."

Oxonica's argument that meaning of the terms arrived at by Hon. Justice Peter Prescott would deprive the words "or Licensed Patents" of any meaning was at first sightly persuasive. However, the definitions clause in the agreement stated that the defined terms would have the meanings given "except where the context otherwise requires." There was a difference between the context of licensing—where it was important to license everything in both the PCT application and all ensuing patents—and the context of royalty payments. In relation to royalty payments, the words "or Licensed Patents" should not be read as applicable because "they made no sense or an unreasonable sense in that context." This was the only way to make rational sense of what Lord Justice Jacob referred to as an "appallingly drafted document."

The appeal court did not take the decision to ignore the words "or Licensed Patent" within the agreement lightly, but felt that such action was required to make rational sense of what was essentially a very badly drafted document.

Accordingly, the judgment was upheld in much the same terms as the original judgment.

5.5.3 Business Consequences of the Judgment

- After the judgment, Oxonica announced that the order and costs would be dealt with at a later hearing.
- An initial amount of £394,000 covering the royalties in dispute was placed in a joint bank account of both parties' solicitors in October 2008.
- An amount of £550,000 was provided for in addition to the £394,000 royalties following the High Court hearing. Oxonica Energy has increased this provision for costs payable to Neuftec arising from this action to £800,000 following consultation with the company's legal advisors.
- Oxonica had cash balances of £1.6 million as of July 7, 2009.
- Oxonica Energy Ltd. was sold to Energenics in September 2009 in full and a final settlement of legal actions between Neuftec and Oxonica was achieved.[33]

5.6 *DSM v. 3D Systems*

Stereolithography is a technology that creates relatively accurate and robust parts rapidly from computer-aided design (CAD) data to aid in product development processes. Further employing stereolithography apparatus (SLA), this technology has been used for functional testing including appearance appraisal, producing multiple replicated parts by means of investment casting, vacuum casting, and other processes to enable functional testing to be performed or for the direct creation of low-volume production parts. This technique has evolved over the past decade to achieve enhanced accuracy and surface finish, and faster production with the development of newer resins to enable tougher materials so that parts that are more robust can be created.

3DS Corporation (3DS) is one of the leading developers of stereolithography machines and materials. DSM Somos has developed a software application that enables lightweight prototypes to be created with enhanced structural integrity.

Based on patented technology from the Milwaukee School of Engineering (MSOE), Tetrashell hollow-build software uses MSOE's Tetralattice technology to facilitate the manufacture of hollow stereolithography parts with variable skin thicknesses, supported by a Tetralattice support structure. Potential application areas include investment casting patterns, reduced-density metal-clad composite structures, and lightweight large, thick-sectioned parts.[34]

3DS was founded in 1986 and is a leading provider of 3D printing, rapid prototyping, and manufacturing systems and parts solutions. Its expertly integrated solutions reduce the time and cost of designing products, and facilitate direct and indirect manufacturing by creating actual parts directly from digital input. These solutions are used for design communication and prototyping as well as for production of functional end-use parts. Its systems utilize patented technologies that create physical objects from digital input. 3DS licenses the complementary 3D Keltool® process, a method for producing steel mold inserts, and currently is developing systems that use composite paste materials for direct manufacturing. In August 2001, 3DS merged with DTM Corp.

DSM Desotech is a global leader in the development and manufacture of ultraviolet light and electron beam (UV/EB) curable materials and operates as a business unit of DSM—a highly integrated group of companies with worldwide interests in life science products, performance materials, and chemicals. DSM reported sales of 8.1 billion euro in 2000 and employs a total workforce of some 22,000 in more than 200 operations throughout the world.

In 1999, DSM Desotech announced its acquisition of Dupont Somos—now DSM Somos. Somos is currently the world's second-largest materials supplier to the rapid prototyping industry, providing stereolithography liquids and selective laser sintering powders used for the creation of 3D models and prototypes directly from digital data. Somos' patented Protofunctional™ materials are used by various industries, including automotive, aerospace, medical, and telecommunications.

On December 31, 2001, DSM Desotech Inc. and 3D Systems Corp. announced the formation of a joint venture, OptoForm LLC, to focus on the development and commercialization of new equipment and materials for rapid tooling and direct and indirect inline manufacturing processes. The materials included ceramics, composite tooling materials, and toughened plastics. This technology was expected to open up new markets aimed at producing parts that are more functional. Research and development activities are taking place in both 3DS's facility in Valencia, California and DSM Desotech's Somos® facility in New Castle, Delaware.

By sharing their mutual expertise, coupled with the strong intellectual property position in this area of both companies, the joint venture would focus on speedy delivery of new products to the marketplace.[35] On January 11, 2006, an in principle agreement was reached between 3DS Corporation and DSM Somos, an unincorporated division of DSM Desotech, Inc. and an affiliate of DSM N.V., to settle the patent litigation pending between the two companies in Germany and to cross-license patents and other intellectual property relating to stereolithography materials.

As part of this settlement, DSM would grant to 3DS a nonexclusive worldwide license outside of Japan to patents relating to stereolithography materials and equipment, and 3DS would grant to DSM a nonexclusive worldwide license outside of Japan to patents and patent applications relating to stereolithography materials and the corresponding right to sell DSM stereolithography materials to users of 3DS's new Viper™ Pro SLA® systems under 3DS's patents, patent applications, and other intellectual property relating to those systems.

The parties contemplate that their current distribution agreement, under which 3DS distributes DSM's stereolithography materials outside of Japan, will continue. While completion of this agreement in principle is subject to the negotiation, execution, and delivery of a definitive agreement, both parties intend to commence commercial activities consistent with the proposed settlement immediately.

"We are pleased to be able to announce this agreement in principle and to begin marketing activities today," said Abe Reichental, 3DS's president and chief executive officer. "We believe that this new arrangement will benefit our stereolithography customers by enabling them to have access to an ever-broadening group of materials for rapid manufacturing and other applications. We are pleased to have been able to negotiate an arrangement which should be beneficial to both 3DS and DSM."

Jim Reitz, Somos® business manager for DSM Desotech, added, "This agreement allows both parties to fully focus on efforts to grow stereolithography technology to the benefit of the rapid prototyping and rapid manufacturing markets."[36]

On March 14, 2008, DSM Desotech Inc. ("Desotech") filed a lawsuit in the United States District Court for the Northern District of Illinois (No. 08 C 153) against 3DS in federal court in Chicago complaining of repeated anticompetitive conduct by 3DS in the market for large-frame SL machines, particularly 3DS's Viper Pro SLA System ("Viper Pro"), and the resin used in those machines, as well as 3DS's infringement of Desotech's intellectual property.[37]

Desotech specifically alleged 3D System

- was guilty of conduct that has violated federal and state antitrust laws and state deceptive trade practice law;
- tortiously interfered with Desotech's prospective business relations;
- infringed two patents owned by Desotech through 3DS's sale of the Viper Pro, SLA® 7000, and SLA® 5000 Systems;
- is improperly conditioning the sale and maintenance of its large-frame SL machines on the purchase of 3DS's own resin for use in those machines, despite the fact that customers would prefer to purchase Desotech's Somos resins for use in their large-frame SL machines.

The strategy of filing of the suit by Desotech was to maintain pressure in the marketplace for a fair competition involving specialty resin used in large-frame SL machines and to keep a lead advantage of its innovations by enforcing its intellectual property rights.

The court handed over a judgment on January 26, 2009. The judge dismissed in part DSM's allegation that 3D's conduct had violated federal and state antitrust laws and state deceptive trade practice law. The Court however directed that discovery may proceed on all outstanding antitrust and patent claims including patent damages issues.[38] The judgment is provided in the appendices at end of this chapter.

On October 27, 2004, Koninklijke DSM N.V. ("DSM") filed a petition in the District Court of Frankfurt, Germany (Case No. 2 06 O 426/04) against 3DS's German subsidiary, 3D Systems GmbH, seeking a preliminary injunction against its sale in Germany of 3D System's Bluestone™ stereolithography resin on the alleged basis that this product infringes German Patent No. DE 69713558 alleged to be jointly owned by DSM, Japan Synthetic Rubber Co. Ltd. (JSR), and Japan Fine Coatings Co. Ltd. (JFC). Accura® Bluestone stereolithography material is an engineered nano-composite resin offered globally for use in 3D Systems' SLA® systems that include the Viper™, iPro™ 8000, iPro™ 8000 MP, iPro™ 9000, and iPro™ 9000 XL. Accura Bluestone material, with its high stiffness and stability, is widely used by Formula 1 race teams to produce scale wind-tunnel models.

On November 25, 2004, 3D Systems GmbH filed an action against DSM, JSR, and JFC in the Federal Patent Court in Munich, Germany (Case No. 4 Ni 58/04) in which it sought a judgment of that Court to invalidate this patent.

On December 13, 2004, DSM withdrew the foregoing petition for a preliminary injunction, but subsequently served on 3D Germany a complaint that was originally filed on November 2, 2004 in the Frankfurt, Germany District Court (Case No. 2-06 0442/04) alleging claims of infringement based on the same patent. Such complaint sought injunctive relief preventing the distribution of Bluestone resin in Germany and an accounting of sales of that product in Germany. 3D Germany filed a reply to DSM's complaint in which, among other things, it requested dismissal of the complaint and assessment of its costs against DSM. 3D Germany also asserted various defenses, including that the Bluestone material does not infringe any valid claim of the patent and that the asserted claims of the patent are invalid. The Frankfurt Court held a hearing on DSM's claims of infringement on May 11, 2005, and on August 8, 2005, the Court issued an opinion ruling in favor of DSM on its claims of infringement. DSM subsequently enforced an injunction issued by the Court against 3D Germany that prohibited its sale of Bluestone resin in Germany and required an accounting of profits on prior sales of that product in Germany. 3D appealed such ruling and complied with such injunction pending the outcome of the appeal and the outcome of 3D Germany's action in the German Federal Patent Court.

Subsequently, the Federal Patent Court in Munich held a hearing on November 8, 2005 on 3D Germany's action against DSM, JSR, and JFC seeking to invalidate this patent and issued a ruling in favor of 3D Systems.[39]

On January 6, 2006, 3D and DSM Desotech, Inc. entered into an agreement in principle to cross-license certain patents and to license certain other technology for the purpose of settling the litigation between the parties. Completion of this arrangement remained subject to the negotiation, execution, and delivery of definitive documentation and other customary conditions.

On January 28, 2010, the German Federal Supreme Court issued its decision invalidating the patent. Because of this decision, the related 2005 decision that DSM obtained in an infringement action before the Frankfurt District Court would stand dismissed based upon this complete revocation of DSM's patent.

After the decision, Reichental, 3D Systems' president and CEO, stated, "3D Systems is very pleased with this decision. As we have said before, we respect the valid intellectual property rights of others. The German Federal Supreme Court's decision to revoke this patent is consistent with our long-held position that DSM's patent was invalid."[40]

5.7 *Veeco Instruments v. Asylum Research*

Asylum Research in Santa Barbara, California is a premier manufacturer of atomic force microscopes including AFMs/ SPMs, for nanoscale science and technology. An AFM/SPM is used for visualizing surfaces and measuring surface properties at the nanometer level.

Veeco Instruments is the world leader in atomic force and scanning probe microscopy, with an installed base of over 8000 systems at university and research/nanotechnology centers worldwide. Veeco makes equipment to develop and manufacture LEDs, solar panels, hard drives, and other devices. It supports its customers through product development, manufacturing, sales, and service sites in the United States, Korea, Taiwan, China, Singapore, Japan, Europe, and other locations.

On September 17, 2003, Veeco Instruments filed a patent infringement suit against Asylum Research in the United States District Court in Los Angeles, California. Veeco claimed that Asylum's MFP-3D atomic force microscope infringes five patents owned by Veeco.[41]

Summary judgment motions were filed in 2004. Asylum asserted that it did not infringe the patents and that the patents are invalid. Asylum also claimed that Veeco patents are invalid for improper inventorship, among other reasons. Veeco requested the court confirm validity of its patents and issue a permanent injunction including monetary damages for willful infringement by Asylum and barring the sale of any infringing Asylum AFMs.

An order was issued on March 19, 2007; the Court issued the order under seal. The Court lifted the seal on November 15, 2007. In this ruling, the Court dismissed two of the five Veeco patents. The Court narrowly interpreted the claims of the two patents Veeco had asserted against Asylum and eliminated 6 of 13 patent claims that Veeco brought against Asylum while all Asylum claims against Veeco remained. The Court found that Asylum did not infringe these two patents. Veeco disagreed with this narrow interpretation on the basis that it had a strong case to appeal this interpretation at the appropriate time.

In November 2007, after considering the arguments in detail, the Court ruled that Veeco had presented sufficient evidence of Asylum's infringement as to three of Veeco's patents to set the case for trial in March 2008 on the remaining issues of the suit concerning the three patents, namely, U.S. Patents Nos. 5,224,376 and 5,237,859 (Atomic Force Microscope) and RE36,488 (Tapping Atomic Force Microscope with Phase or Frequency Detection) and assertion by Asylum that Veeco's patents are invalid and unenforceable and a counterclaim for infringement of a patent licensed by Asylum as part of the lawsuit.[42]

Co-founder and CEO of Asylum Research Jason Cleveland commented,

> Four years ago Veeco chose to pursue litigation rather than innovation. Their obvious hope was to financially starve a young competitor. That strategy has backfired and they now find themselves with a damaged portfolio and facing a possible infringement ruling for using our technology—technology that enables much faster, lower noise AFMs that will be required for Veeco to be competitive in the future. While we are obviously very happy with this victory, we look forward to finishing the job at trial.[43]

A date-wise recording of the court proceedings toward the end of the trial illustrates how the two parties finally agreed to mutually settle the matter.[44]

08-05-2008	MINUTES OF Status Conference re Jury Trial held before Judge George H Wu: The parties indicate that the case has tentatively settled. The Court orders the parties to return for a Status Conference re Settlement on 8/6/2008 at 9:30 AM. Court Reporter: Wil Wilcox. (jp) (Entered: 08/06/2008)
08-06-2008	MINUTES OF Status Conference held before Judge George H Wu: Parties inform the Court that the settlement has not been finalized. The Status Conference re Settlement is continued to August 9, 2008 at 10:00 a.m. Court Reporter: Wil Wilcox. (pj) (Entered: 08/07/2008)
08-08-2008	MINUTES OF Status Conference Re: Settlement held before Judge George H. Wu. Status Conference is held off the record. The parties have finalized a settlement. The dismissal in this matter shall be filed by no later than noon on 8/11/2008. Court Reporter: N/A. (rj) (Entered: 08/11/2008)
08-11-2008	STIPULATION for Judgment as to Dismissal filed by plaintiffs Veeco Instruments Inc., Veeco Metrology LLC. (Barquist, Charles) (Entered: 08/11/2008)
	ORDER RE JUDGMENT
08-14-2008	ORDER RE JUDGMENT by Judge George H Wu, This action, including all claims, defenses, affirmative defenses, and counterclaims, shall be dismissed with prejudice except insofar as this Court shall retain exclusive, continuing jurisdiction to enforce, administer, and/or interpret this Stipulated Judgment and Order, and the Agreement. Veeco and Asylum shall each bear their own legal costs and attorneys' fees incurred in this action. [510] (MD JS-6, Case Terminated). (pj) (Entered: 08/15/2008)

Under the terms of the settlement, Asylum and Veeco agreed to drop all pending claims against each other and agreed to a five-year, worldwide cross-license of each other's patents and a mutual covenant not to sue on patents either party has a right to assert. Asylum will pay an initial license fee to Veeco plus an ongoing royalty for the five-year term of the cross-license. As part of the settlement, Asylum acknowledged the validity of the Veeco patents asserted in the case. Other terms of the agreement were not disclosed.

Jason Cleveland, CEO of Asylum, said,

> We are pleased that this litigation is in our past and that we can now move forward. Asylum spends a significant percentage of our revenues on research and development and that is reflected in our fast growing patent portfolio. The cross-licensing of portfolios will allow both us and Veeco to bring better products to market which is good news for customers.

Roger Proksch, president of Asylum, added, "The nanotechnology market values innovation. Everyone in the field will welcome the end to this struggle because it allows us to return to doing what we do best—making great AFMs and supporting our loyal customers."[45]

In another proceeding at the European Patent Office (EPO), Munich, Germany, Asylum Research had opposed Veeco's Patent No. 839,312 (the 312 patent), involving the use of an AFM in tapping mode with phase or frequency detection to image the topography and surface characteristics of a sample. However, in January 2007, the EPO ruled in favor of Veeco Instruments, Inc. and dismissed the opposition filed by Asylum Research Inc.[46]

5.7.1 Postjudgment Proceedings

On October 8, 2010, Bruker Corporation announced the closing of its acquisition of the Atomic Force Microscopy (AFM) and the Optical Industrial Metrology (OIM) instruments businesses from Veeco Instruments, Inc.[47]

With this acquisition Veeco Instruments, the industry-leading AFM scientific instruments business headquartered in Santa Barbara, California, as well as the OIM business based in Tucson, Arizona, along with the global AFM/OIM field sales, applications, and support organizations, became part of the Bruker Nano Division, which is part of the Bruker AXS Group, adding more than 350 employees in 11 countries. The 2009 revenues for the AFM and OIM instruments businesses were derived approximately 38% from Asia-Pacific, 31% from the Americas, and 31% from Europe, and in terms of customers, approximately 45% from applied/industrial customers and 55% from academic/government customers.

The acquired AFM and OIM businesses are highly complementary to Bruker's existing systems and solutions, and the combined product portfolio transforms Bruker into a global leader in materials research and nanotechnology analysis instrumentation for surface analysis in materials, life science, and nanotechnology R&D and quality control.

Frank Laukien, president and CEO of Bruker Corporation, stated,

> We are very excited about the addition of these highly regarded AFM and OIM businesses to Bruker, as they complement our focused product and market strategies very well. With these additional high-performance and industry-leading products, Bruker can now serve its global customers and markets even better. Moreover, we cordially welcome the many talented and motivated new AFM and OIM colleagues who have just joined Bruker.

Dr. R. Munch, president of Bruker Nano, Inc. commented:

> Together with Bruker, we now have a tremendous new ability to further develop innovative products that will evolve the industry and how we measure and obtain nanoscale information. Bruker has been extremely supportive from the start and is dedicated to ensuring that our current and future customers receive the highest performing and most innovative instruments with unsurpassed service.

In 2011, the acquired Bruker Nano AFM and OIM businesses were forecasted to contribute revenues greater than $130 million, and adjusted operating margins (excluding acquisition-related and noncash intangible amortization charges) of greater than 15%.

5.8 *Modumetal, Inc. v. Integran, Inc.*

On October 4, 2010, Seattle-based Modumetal, Inc. (Modumetal) filed a complaint against Integran Technologies in the United States District Court, Western District of Washington in Seattle, Washington (No. 2:10-CV-01592) seeking declaratory judgment of patent invalidity and noninfringement of U.S. Patent Nos. 5,433,797, 7,320,832, or 7,553,553 owned by Integran Technologies, Inc.[48]

On August 24, 2010, Integran sent a letter to the president and CEO of Modumetal indicating that Integran believed that Modumetal might be infringing valid patent rights owned by Integran. In particular, Integran claimed in the letter to own at least four issued U.S. patents and four pending and published applications.

On September 28, 2010, Integran specifically referred Modumetal to Claim 6 of U.S. Patent No. 5,433,797; Claim 1 of U.S. Patent No. 7,553,553; and Claim 1 of U.S. Patent No. 7,320,832 (collectively herein after the "asserted claims") as claims that are relevant to Modumetal's technology.

On March 15, 2011, Integran Technologies, Inc. filed claims for patent infringement against Modumetal, Inc. in the United States District Court, Western District of Washington in Seattle, Washington.

In the lawsuit filed in Seattle, Integran asserted that Modumetal was infringing or had infringed U.S. Patent Nos. 5,433,797; 7,320,832; 7,553,553; 5,352,266; and 7,824,774 owned by Integran seeking permanently to enjoin Modumetal from further sales or use of Modumetal's infringing nanolaminate materials and technology. Integran also requested that the Court compensate Integran for Modumetal's infringement of Integran's patents.[49]

On August 8, 2011, Integran, Inc. announced that Integran and Modumetal, Inc. (Modumetal) had signed a binding settlement agreement to end litigation. All terms of the agreement had been finalized and the patent infringement litigation against Modumetal was dropped by Integran. The agreement eliminates the need for any further court proceedings or decisions relating to damages or injunctive relief.

Under the terms of the settlement, the parties came to an agreement regarding Modumetal's research and commercial activities relating to its nanolaminate technology. The parties have agreed that details of the agreement, including financial terms of the settlement, will be kept confidential.

5.9 *MTS Systems Corp. v. Hysitron, Inc.*

On March 14, 2007, Hysitron, Inc. filed a suit in the United States District Court, District of Minnesota (CIVIL NO. 07-1533 [ADM/AJB]) with Jury demand, alleging MTS Systems Corporation of infringing (directly infringed, induced infringement, and contributed to the infringement) U.S. Patent Nos. 6,026,677 and 5,553,486 entitled "Apparatus for Microindentation Hardness Testing and Surface Imaging Incorporating a Multi-Plate Capacitor System."

The compliant stated that the Defendant had infringed the said patents through the manufacture, use, sale, and offer for sale of indentation testing devices with a scanned probe microscope apparatus, including NANOVision™, as well as Defendant's manufacture, use, sale, and offer for sale of scanned probe microscope apparatuses, including NANOVision, to be used with indentation testing device.[50]

All the dockets and filings related to this case may be viewed at http://dockets.justia.com/docket/minnesota/mndce/0:2007cv01533/90143/.

> **ORDER OPINION** by Magistrate Judge Arthur J. Boylan of the United States District Court dated January 4 2010, stated, "Based upon the record, memoranda, and oral arguments of counsel, **IT IS HEREBY ORDERED** that Plaintiff's Motion to Exclude [Docket Nos. 181] is **GRANTED IN PART** and **DENIED IN PART**, and Plaintiff's Motion to Compel [Docket No. 186] is **GRANTED** as set forth herein and as follows:
>
> - On or before February 1, 2010, Defendant shall pay to Plaintiff the sum of five thousand dollars ($5,000.00) as a sanction for failing to timely supplement its motion and as reasonable fees associated with Plaintiff bringing the present motion.
> - On or before January 15, 2010, Defendant shall provide Plaintiff with a Third Supplemental Prior Art Statement that adheres to the requirements of the Pretrial Scheduling Order.
> - On or before February 1, 2010, Plaintiff will submit to this Court a proposed Sixth Amended Scheduling Order providing a timeline and number of interrogatories and depositions needed for Plaintiff to conduct discovery related to Defendant's invalidity claims.
> - On or before February 1, 2010, Defendant must produce nonprivileged information responsive to Plaintiff's Document Request No. 15, including, but not limited to, testing reports for the accused device."[50]

On July 23, 2010, the parties Hysitron Incorporated and MTS Systems Corporation executed a settlement agreement ("the Agreement"). The Agreement dismissed all claims and counterclaims in the patent infringement lawsuit between the two companies. Pursuant to the agreement, MTS will pay Hysitron $7.5 million and neither party admits any liability or wrongdoing.[51]

5.10 *EV Group v. 3M*

On July 11, 2008, EV Group, a leading supplier of wafer-bonding and lithography equipment for the advanced semiconductor and packaging, MEMS, silicon-on-insulator (SOI), and emerging nanotechnology markets, filed a patent infringement lawsuit in New York Southern District Court against the 3M Company ("3M") (case no.1:2008cv06287). The complaint alleged that 3M had infringed upon a U.S. Patent through the marketing and sale of the 3M Wafer Support System, which is used in the production of silicon wafers. The complaint seeks damages to compensate EVG for 3M's wrongful infringement and an injunction against 3M from all future infringement of the patent.

On December 22, 2008, 3M and EV Group (EVG) agreed to settle this patent infringement litigation.

Under the terms of the settlement, the details of which are confidential, 3M, its customers, and 3M's licensed suppliers of 3M's Wafer Support System will continue to make, sell, and use the Wafer Support System in global semiconductor and packaging markets. EVG will continue to defend its patent portfolio and protect its intellectual property.[52]

A summary of the case as recorded shows the swift manner in which the matter was settled by the two parties. This may serve as a model for business-led settlements with regard to nanotechnology patent litigations in the future.

Date	#	Docket Text
12/5/2008	5	NOTICE OF VOLUNTARY DISMISSAL Pursuant to Rule 41(a)(1)(A)(i) of the Federal Rules of Civil Procedure, the plaintiff(s) and or their counsel(s), hereby give notice that the above-captioned action is voluntarily dismissed without prejudice against the defendant(s) 3M Company. So Ordered (Signed by Judge Naomi Reice Buchwald on 12/4/08) (js) (Entered: 12/05/2008)
12/4/2008		Mailed notice to Commissioner of Patents and Trademarks to report the termination of this action. (dt) (Entered: 12/04/2008)
12/3/2008	4	NOTICE OF VOLUNTARY DISMISSAL Pursuant to Rule 41(a)(1)(A)(i) of the Federal Rules of Civil Procedure, the plaintiff(s) and or their counsel(s), hereby give notice that the above-captioned action is voluntarily dismissed, against the defendant(s) 3M Company. Document filed by all plaintiffs. (Chin, Steven) (Entered: 12/03/2008)
11/6/2008	3	STIPULATION AND ORDER It is hereby stipulated and agreed that plaintiffs' deadline to serve the summons and complaint in the above-identified action shall be extended for 30 days, from November 10, 2008 to and including December 10, 2008. The parties are usually engaged in settlement discussions and this extension will provide the parties time to seek a resolution of the dispute. (Signed by Judge Naomi Reice Buchwald on 11/5/08) (mme) (Entered: 11/06/2008)
7/14/2008		***NOTE TO ATTORNEY TO E-MAIL PDF. Note to Attorney Peter Jonathan Toren for noncompliance with Section (3) of the S.D.N.Y. 3rd Amended Instructions For Filing An Electronic Case or Appeal and Section 1(d) of the S.D.N.Y. Procedures For Electronic Case Filing. E-MAIL the PDF for Document 1 Complaint, 2 Rule 7.1 Corporate Disclosure Statement to: case_openings@nysd.uscourts.gov. (jeh) (Entered: 07/14/2008)

7/11/2008	2	RULE 7.1 CORPORATE DISCLOSURE STATEMENT. No Corporate Parent. Document filed by EV Group E. Thallner GmbH, Erich Thallner. (jeh) (Entered: 07/14/2008)
7/11/2008	1	COMPLAINT against 3M Company. (Filing Fee $350.00, Receipt Number 656533) Document filed by EV Group E. Thallner GmbH, Erich Thallner. (jeh) (Entered: 07/14/2008)
7/11/2008		SUMMONS ISSUED as to 3M Company. (jeh) (Entered: 07/14/2008)
7/11/2008		Case Designated ECF. (jeh) (Entered: 07/14/2008)
7/11/2008		Magistrate Judge Debra Freeman is so designated. (jeh) (Entered: 07/14/2008)

5.11 *Tekmira v. Alnylam*

On March 16, 2011, Tekmira Pharmaceuticals Corp. (TPC) and Protiva Biotherapeutics, Inc. (Provita), collectively called "Tekmira", filed a legal complaint in the Business Litigation Session of the Massachusetts Superior Court against Alnylam Pharmaceuticals for preliminary and permanent injunctive relief, an equitable accounting, a constructive trust, damages, and other relief arising out of Alnylam's misappropriation of Tekmira's confidential and proprietary information and trade secrets, unfair and deceptive trade practices, unjust enrichment, and unfair competition. For the full compliant, refer to Endnote 53.

Tekmira is a leading developer of lipid nanoparticle delivery technologies that are thought to help get the RNA-silencing therapies where they need to go in cells. Tekmira has spent more than 500 person-years of effort and over $200 million developing proprietary and confidential novel lipids and formulations, as well as proprietary techniques for large-scale manufacturing of its formulations. The knowledge and information gained by Tekmira through its years of formulation and manufacturing work are all part of Tekmira's siRNA delivery technology. Tekmira has taken all steps to protect its confidential and proprietary information and trade secrets. Tekmira's confidential and proprietary information and trade secrets derive value from not being generally known to the public or to others who can obtain value from their disclosure or use. Tekmira expended considerable sums of money and time developing its confidential and proprietary information and trade secrets.

Alnylam began collaborating with TPC and Protiva before they merged in March 2008, and continued collaborating with Tekmira after the merger. In the course of the collaboration relationships, Tekmira made a number of disclosures of its delivery technology to Alnylam. Tekmira did so under the protection of written agreements that restricted Alnylam's right to use Tekmira's confidential information and trade secrets and that strictly prohibited Alnylam from disclosing such information to third parties without first obtaining Tekmira's consent. Tekmira did not grant Alnylam ownership of its delivery technology; the delivery technology remained Tekmira's property before, during, and after Tekmira's collaboration with Alnylam.

Tekmira claims that in a carefully orchestrated series of wrongful acts, Alnylam took advantage of its confidential relationship as a collaborator to gain access to and exploit for its own benefit some of Tekmira's most valuable and highly confidential technology, for purposes and activities that were not authorized by Tekmira. Alnylam abused its confidential status repeatedly, improperly using Tekmira's technology to derive additional formulations for Alnylam's own benefit. Those purported Alnylam formulations actually contain, are based on, and are in whole or in part developed from Tekmira's technology.

In its complaint, Tekmira alleges that Alnylam engaged and continues to engage in at least the following wrongful conduct without Tekmira's authorization or consent:

- Alnylam improperly used Tekmira's "Lead Formulation," which Alnylam used for its first VSP product, to develop derivative formulations for its own benefit.

- Alnylam applied for patents in which it wrongfully claimed to own and improperly disclosed Tekmira's Lead Formulation and the wrongfully derived derivatives, including without limitation in provisional patent applications 61/242,783 and 61/148,366, and PCT patent applications US2009/036223 and US2009/061381.

- Alnylam improperly used and disclosed other Tekmira formulations to develop derivative formulations, including but not limited to Tekmira's MC3 formulation.

- Alnylam applied for patents in which it wrongfully claimed to own and improperly disclosed Tekmira's formulations and the wrongfully derived derivatives such as MC3, including without limitation in provisional patent applications 61/154,350, 61/171,439, 61/185,438, and PCT patent applications US2009/063927, US2009/063931, and US2009/063933.

- Alnylam improperly disclosed portions of Tekmira's secret step-by-step formulation manufacturing instructions to at least one of its third-party collaborators.

- Alnylam improperly used and disclosed in patent filings certain highly confidential Tekmira information that Alnylam had obtained confidentially for use in its regulatory filings.

- Alnylam is falsely representing to current and potential industry partners that it invented, developed, and owns formulation technology that it stole from Tekmira.

Alnylam stated that since 2007, Alnylam has provided over $45 million in funding, including equity purchases, to Tekmira and provided approximately

$6 million in manufacturing-related funding in 2010 with similar levels expected in 2011. In exchange, Alnylam has obtained broad license rights to Tekmira intellectual property in addition to exclusive rights, with the sole right to sublicense, to many patents and patent applications including the so-called "Semple" and "Wheeler" patent families (Semple U.S. Patent No. 6,858,225 and Wheeler U.S. Patent Nos. 5,976,567 and 6,815,432). In addition, Alnylam maintains exclusive research and discovery collaborations on novel lipid nanoparticle (LNP) formulations through agreements with AlCana Technologies, Inc., the University of British Columbia, and the Massachusetts Institute of Technology

Tekmira hopes for the following relief:

- Judgment in Tekmira's favor on each count;
- Compensatory damages;
- Exemplary and enhanced damages;
- An award directing Alnylam to disgorge to Tekmira all monies and/ or profits derived from the wrongful conduct alleged herein;
- An award to Tekmira of the amount by which Alnylam has been unjustly enriched;
- Reasonable royalties for Alnylam's improper use of Tekmira's technology;
- A preliminary and permanent injunction enjoining and restraining Alnylam, and its officers, directors, agents, servants, employees, attorneys, and all others acting under, by, or through them, directly or indirectly, from improperly possessing, obtaining, transferring, using, or disclosing any Tekmira confidential and proprietary information or trade secrets, including any Alnylam material, products, or technology that wrongfully contain, are based on, and/or are derived in whole or in part from any of Tekmira's confidential and proprietary information or trade secrets;
- A preliminary and permanent injunction enjoining and restraining Alnylam, and its officers, directors, agents, servants, employees, attorneys, and all others acting under, by, or through them, directly or indirectly, from making, publishing, disseminating, circulating, or placing before the public statements in which Alnylam claims ownership of siRNA delivery technology that wrongfully contains, is based on, and/or is derived in whole or in part from any Tekmira confidential and proprietary information and/or trade secrets;
- An accounting of any monetary or other benefits received by Alnylam as a result of its wrongful conduct;
- A constructive trust over all information, patent applications, patents, technology, products, and other materials in the possession, custody, or control of Alnylam that wrongfully constitute, contain,

were based on, and/or derived in whole or in part from the use of
Tekmira's confidential and proprietary information and/or trade
secrets, and an order that Alnylam immediately transfer to Tekmira
all rights, title, and interest in such information, patent applications,
patents, materials, technology, and products;

- Prejudgment interest according to proof;
- Reasonable attorneys' fees and costs of suit; and
- Such other relief that the Court deems just and proper.

In May 2011, the BLS Court put in place an order to protect the parties'
confidential information during the litigation and ruled that the confidential documents or information produced in this Tekmira–Alnylam litigation
could not be used in the patent interference proceeding currently ongoing
between Tekmira and Alnylam.

On June 11, 2011, Tekmira filed an amended complaint against
Alnylam Pharmaceuticals, Inc. with the Business Litigation Session of the
Massachusetts Superior Court. Tekmira's revised complaint brings new
claims alleging violation of contract, breach of the intended covenant of good
faith and fair dealing, tortious interference with contractual associations,
and civil conspiracy. The amended complaint additionally adds AlCana
Technologies as a defendant and asserts claims alleging misappropriation of trade secrets, tortious interference with contractual relations, unfair
enrichment, unjust and misleading acts and trade methods, and civil conspiracy against AlCana. Throughout the litigation, Alnylam has stated that,
among other claims, Tekmira is in violation of the license and production
agreements involving the two companies.

As referred to in the amended complaint, Tekmira is seeking relief in the
form of damages, including the royalties and profits Alnylam and AlCana
would receive from the alleged improper use of Tekmira's technology and
the termination of Alnylam's license to Tekmira's technology.[54]

The United States Patent and Trademark Office has declared a patent inter-
ference (No. 105,792) to determine priority to subject matter of Alnylam's U.S.
Patent No. 7,718,629 in light of Tekmira's U.S. Patent Application 11/807,872
claiming the EG5 siRNA sequence used in Alnylam's ALN-VSP product.
Tekmira believes certain claims in Alnylam's 629 patent are invalid and that
Tekmira filed on the claimed sequence prior to Alnylam.

In September 2011, the BLS Court dismissed Alnylam's counterclaim of
defamation against Tekmira. Alnylam accused Tekmira of defamation for
public statements made upon the filing of the lawsuit in March 2011. The BLS
Court ruled that the law permitted Tekmira to communicate with its shareholders as it did upon the filing of the lawsuit. According to Massachusetts
law, Tekmira is entitled to an award of reasonable attorneys' fees and costs.

Discovery is currently underway, and Tekmira has requested that a trial
date be set for the fall of 2012.

5.12 Other Nano Patents Litigation Snippets

KLA-Tencor Corp. was formed in April 1997 through the merger of KLA Instruments (KLA) and Tencor Instruments (Tencor), two long-time leaders in the semiconductor equipment industry. KLA-Tencor Corp. is an advanced measurement system developer designing tools to enable chipmakers to identify problems early in the manufacturing process.

Xitronix Corporation, founded in 2006, is the leading supplier of photo-modulated reflectance (PMR) metrology equipment to the worldwide semiconductor manufacturing market. PMR technology provides process engineers the unprecedented capability to measure the electrical and material properties in nanoscale semiconductor structures during the manufacturing process. The direct sensitivity to active electronic properties of nanostructures provided by PMR is a high-priority need in chip manufacturing now and into the foreseeable future. Other available characterization technologies do not provide direct sensitivity to electronic properties and are not effective for process control. Xitronix has numerous patents and patent applications based on its state-of-the-art PMR technology.

KLA-Tenor Corp filed a suit in the Federal Court in Austin, Texas on September 24, 2008 (Case No. A-08-CA-723-SS), seeking a permanent injunction against Xitronix for allegedly using its patented technology. The trial in this case began on November 1, 2010 and concluded with a jury verdict on November 5, 2010. The jury returned a verdict finding Xitronix had infringed Claim 7 of U.S. Patent 7362441 (441 filed on September 13, 2006 and granted on September 22, 2008) titled "Modulated Reflectance Measurement System Using UV Probe" but had not infringed any other claims. The jury also found, however, that Claim 7 of the 441 patent was invalid as anticipated by prior art, namely the 611 patent and the Therma-Probe device. Further, the jury found all of the asserted claims 7, 9, 11, and 12 of the 441 patent were invalid due to obviousness. The judgment was signed on January 31, 2011.[55]

The patent in question applied to Xitronix's nondestructive PMR solution, which is meant to enable faster, more cost-effective process control measurements to control the material and electronic properties of semiconductor nanostructures. Xitronix developed the technology in response to the semiconductor industry's focus on engineering properties as chips became increasingly smaller and reached physical limits.

In February 2011, a jury ruled KLA-Tencor Corp.'s patent claims as invalid and that Xitronix did not infringe.

This favorable judgment now enables Xitronix to provide its unique process control solution to the semiconductor industry with a clear competitive advantage as the sole source provider for the technology to leading-edge manufacturers. Although the patent infringement suit delayed its market operations, its postlitigation strategy is to focus on key early adopters who understand the competitive advantage afforded by its technology.[56]

On April 10, 2001, KLA Tencor Corp. and Therma-Wave, Inc. announced the settlement of their pending litigation. As part of the settlement, KLA-Tencor granted Therma-Wave a license on its U.S. Patent No. 4,899,055 entitled "Thin Film Thickness Measuring Method." In exchange, Therma-Wave granted KLA-Tencor a license on its U.S. Patent No. 5,596,406, entitled "Sample Characteristic Analysis Utilizing Multi Wavelength and Multi Angle Polarization and Magnitude Change Detection" and U.S. Patent Nos. 5,798,837 and 5,900,939, both entitled "Thin Film Optical Measurement System and Method with Calibrating Ellipsometer." In addition, Therma-Wave agreed to modify its products to avoid using certain technology patented by KLA-Tencor and both parties agreed to a moratorium on any patent litigation for an undisclosed period of time. The settlement also resulted in a payment of an undisclosed sum to KLA-Tencor.[57]

On May 25, 2007, KLA-Tencor Corporation completed its acquisition of Therma-Wave Corporation.[58]

Nanometrics Inc. v. KLA-Tencor Corp: On August 3, 2011, KLA-Tencor Corp. (KLAC), a maker of computer chip fabrication equipment, was sued in Federal Court in Delaware by rival Nanometrics Inc., (court case number 1:11-cv-00685-MMB), which accused it of infringing two of their U.S. patents 6,982,793 B1 titled "Method and Apparatus for Using An Alignment Target with Designed In Offset" and 7,230,705 B1 titled "Alignment Target with Designed Offset" and seeks treble damages for willful infringement after a jury trial. Nanometrics is and has been the owner by assignment of all rights, title, and interest in and to the said patents.[59]

The case was assigned to Judge Michael M. Baylson on August 10, 2011.

On June 28, 2010, Canon USA filed a complaint for infringement of Canon's U.S. Patent Nos. 5,903,803 and 6,128,454, against the following 35 Respondents concerning certain toner cartridges and their photosensitive drums sold for use in Canon or Hewlett Packard laser beam printers. The respondents are Ninestar Image Int'l Ltd.; Ninestar Technology Co., Ltd.; Ninestar Management Co., Ltd.; Zhuhai Seine Technology Co., Ltd.; Seine Image Int'l Co., Ltd.; Ninestar Image Co., Ltd.; Ziprint Image Corp.; Nano Pacific Corp.; Ninestar Tech. Co., Ltd.; Town Sky, Inc.; ACM Technologies, Inc.; LD Products, Inc.; Printer Essentials.com, Inc.; Ninestar Technology Co., Ltd.; Ninestar Management Co., Ltd.; Zhuhai Seine Technology Co., Ltd.; Seine Image Int'l Co., Ltd.; Ninestar Image Co., Ltd.; Ziprint Image Corp.; Nano Pacific Corp.; Ninestar Tech. Co., Ltd.; Town Sky, Inc.; ACM Technologies, Inc.; LD Products, Inc.; Printer Essentials.com, Inc.; XSE Group, Inc.; Image Star; EIS Office Solutions, Inc.; 123 Refills, Inc.; Copy Technologies, Inc.; Compu-Imaging, Inc.; Red Powers, Inc.; LaptopTraveller.com; Direct Billing International, Inc.; and OfficeSupplyOutfitters.com.

Canon initiated its proceeding to obtain an Order from the ITC prohibiting the Respondents from importing into the United States and selling in the United States the accused toner cartridges and their photosensitive drums.

On April 18, 2011, Canon USA announced that it is pleased and satisfied with the agreed upon Consent Order that will terminate its U.S. International Trade Commission (ITC) proceeding against these 35 Respondents, that Consent Order, which was approved by the Administrative Law Judge in charge of the proceeding on April 8, 2011 and is expected to be entered by the ITC approximately 30 days thereafter, acknowledges the validity of the involved patent claims, and acknowledges that the Respondents may not import into the United States or sell in the United States the accused toner cartridges and their photosensitive drums without a license or consent from Canon. Canon announced that it has no present intention to grant a license to the Respondents.

Endnotes

1. Sichko, A. The business review, July 6 2009, http://www.bizjournals.com/albany/stories/2009/07/06/daily3.html.
2. Sichko, A. The business review, April 12, 2010, http://www.bizjournals.com/albany/stories/2010/04/12/story6.html.
3. http://greenpatentblog.com/__oneclick_uploads/2009/05/nanosys-complaint.pdf
4. Nanosys reaches settlement of patent infringement lawsuit against Nanoco, July 23, 2009, http://www.nanosysinc.com/in_the_news/nanosys-reaches-settlement-of-patent-infringement-lawsuite-against-nanoco-technologies-for-quantum-dot-technology/
5. Nanoco settles patent infringement lawsuit with Nanosys, Inc. for quantum dot technology, July 23, 2009, http://www.nanocotechnologies.com/Latest/Nanoco_Settles_Patent_Infringement_Lawsuit_with_Nanosys_Inc_for_Quantum_Dot_Technology/134.aspx
6. http://media.corporate-ir.net/media_files/irol/12/120920/Markman_Order_072508.pdf
7. http://media.corporate-ir.net/media_files/irol/12/120920/Summary_Judgment_Order-redacted-111609.pdf
8. http://www.prnewswire.com/news-releases/dupont-air-products-nanomaterials-llc-receives-ruling-in-patent-invaliditynon-infringement-case-against-cabot-microelectronics-73676832.html.
9. http://google.brand.edgar-online.com/EFX_dll/EDGARpro.dll?FetchFilingHtmlSection1?SectionID=7247282-35438-71543&SessionID=y7UJHe-yRPC6Qs7.
10. Cabot Microelectronics Corporation, Cabot Microelectronics' patents upheld in patent enforcement action against DuPont Air Products NanoMaterials, LLC, July 8, 2010, 0 http://www.globenewswire.com/newsroom/news.html?d=196034
11. Cabot Microelectronics' patent infringement litigation against DuPont Air Products NanoMaterials, November 25, 2009, http://www.azonano.com/news.aspx?newsID=14832
12. *NANO-Proprietary, Inc. v. Canon Inc.*, No. A-05-CA-258-SS, slip op. at 10-11 (W.D.Tex. Nov. 14, 2006).

13. *AN-Proprietary Inc. v. Canon Inc.*, No. A-05-CA-258-SS, 2007 WL 628792, at *3-14 (W.D.Tex. Feb. 22, 2007).
14. *AN-Proprietary, Inc. v. Canon Inc.*, No. A-05-CA-258-SS, slip op. at 3 n.1, 2007 WL 1516793 (W.D.Tex. May 3, 2007).
15. *Nano-Proprietary, Inc. v. Canon, Inc. USA*, http://caselaw.findlaw.com/us-5th-circuit/1057494.html
16. Richards, D. June 3, 2007, http://www.smarthouse.com.au/TVs_And_Large_Display/SED/X2X4P5T7?print=1.
17. Harding, R., Canon clear to launch new type of TV, *Financial Times*, December 2, 2008.
18. http://www.ft.com/cms/s/0/577ed3f0-c011-11dd-9222-0000779fd18c.html
19. Wikipedia, Surface-conduction electron-emitter display, http://en.wikipedia.org/wiki/Surface-conduction_electron-emitter_display
20. http://www.sed-tv-reviews.com/
21. http://www.linkedin.com/company/abraxis-bioscience
22. http://www.findforms.com/pdf_files/ded/36850/202.pdf
23. http://www.findforms.com/pdf_files/ded/36850/655.pdf
24. http://www.findforms.com/pdf_files/ded/36850/555.pdf
25. http://www.findforms.com/pdf_files/ded/36850/202.pdf
26. http://www.findforms.com/pdf_files/ded/36850/491.pdf
27. http://www.findforms.com/pdf_files/ded/36850/546.pdf
28. Prendergast, W.F. and Schafer, H.N., Nanocrystalline pharmaceutical patent litigation: The first case, *Nanotechnology Law & Business*, 5(2), 157–162, 2008.
29. http://www.findforms.com/single_form.php/form/127854/Judgment_District_Court_of_Delaware_District_Court_of_Delaware_Delaware
30. Neutral Citation Number: [2008] EWHC 2127 (Pat); Case No: HC 07 C 00437 in the High Court of Justice Chancery Division Patent Court.
31. Reporting Judge: [2009] EWCA Civ 668 Neutral Citation Number: [2009] EWCA Civ 668 ; Case No: A3/2008/2531.
32. http://www.andlaw.eu/blog_detail.php?news_id=3.
33. http://www.oxonica.com/about/about_company.php
34. Engineer Live, Creating new opportunities to help cut costs with stereolithography, http://www.engineerlive.com/Design-Engineer/Time_Compression/Creating_new_opportunities_to_help_cut_costs_with_stereolithography/21566
35. http://www.3dsystems.com/newsevents/newsreleases/pr-Dec_19_01.asp
36. http://www.3dsystems.com/newsevents/newsreleases/pr-Jan_11_2006.asp
37. DSM Desotech, Inc. files antitrust and patent infringement lawsuit against 3D Systems Corporation, March 31, 2008, http://www.dsm.com/nl_NL/html/dsmd/news_items/news_item25.htm
38. http://docs.justia.com/cases/federal/district-courts/illinois/ilndce/1:2011cv06356/259901/1/0.pdf?1316122962
39. http://www.wikinvest.com/stock/3D_Systems_(TDSC)/Koninklijke%20Dsm%20System%20Gmbh
40. http://www.3dsystems.com/press-releases/3d-systems-announces-revocation-german-federal-supreme-court-dsm-german-patent
41. *Veeco Instruments Inc., et al. v. Asylum Research Corporation*, CV 03-6682 SVW (USDC Central District California).

42. Asylum research eliminates half of Veeco patent claims, November 28, 2007, http://news.thomasnet.com/companystory/Asylum-Research-Eliminates-Half-of-Veeco-Patent-Claims-806951#_normalStart

43. http://www.AsylumResearch.com

44. http://www.legalmetric.com/cases/patent/cacd/cacd_203cv06682.html

45. http://www.bloomberg.com/apps/news?pid=newsarchive&sid=alxBo5uEQlok

46. Asylum Appeals Veeco Patent Decision, http://www.photonics.com/Article.aspx?AID=28544

47. Bruker completes acquisition of the atomic force microscopy and optical industrial metrology instruments businesses from Veeco, October 7, 2010, http://www.brukeraxs.com/news_article.html?&tx_ttnews[tt_news]=213&cHash=bed68327b1&utm_source=adwords&utm_medium=cpc_USA&utm_content=veeco_afm_is_now_bruker&utm_campaign

48. http://www.legalmetric.com/complaints/wawd%202-10cv01592-1-1.pdf

49. http://www.andhranews.net/Business/2011/Integran-Files-Patent-Infringement-Claims-Against-10668.htm

50. http://docs.justia.com/cases/federal/district-courts/minnesota/mndce/0:2007cv01533/90143/126/0.pdf?ts=1218543095

51. http://agreements.realdealdocs.com/Settlement-Agreement/SETTLEMENT-AGREEMENT

52. http://www.evgroup.com/en/about/news/44602/

53. http://files.shareholder.com/downloads/ABEA-50QJTB/0x0x452042/58dbf0f3-0a5a-4798-8710-d5fa9a975dc7/Complaint.pdf

54. Tekmira Pharmaceuticals files amended complaint in Alnylam lawsuit, June 5, 2011, http://wallstreetpit.com/77260-tekmira-pharmaceuticals-files-amended-complaint-in-alnylam-lawsuit

55. http://law.justia.com/cases/federal/district-courts/texas/txwdce/1:2008

56. Xitronix wins patent infringement case against KLA-Tencor, February 8, 2011, http://www.nanowerk.com/news/newsid=20066.php

57. KLA-Tencor and Therma-Wave settle lawsuits, April 10, 2001, http://www.kla-tencor.com/corporate-releases/kla-tencor-and-therma-wave-settle-lawsuits.html

58. KLA-Tencor completes acquisition of Therma-Wave, May 25, 2007, http://www.kla-tencor.com/corporate-releases/kla-tencor-completes-acquisition-of-therma-wave.html

59. http://morrisjames.files.wordpress.com/2011/08/nanometrics-incorporated-v-kla-tencor-corporation.pdf

Appendices

Judgement Orders

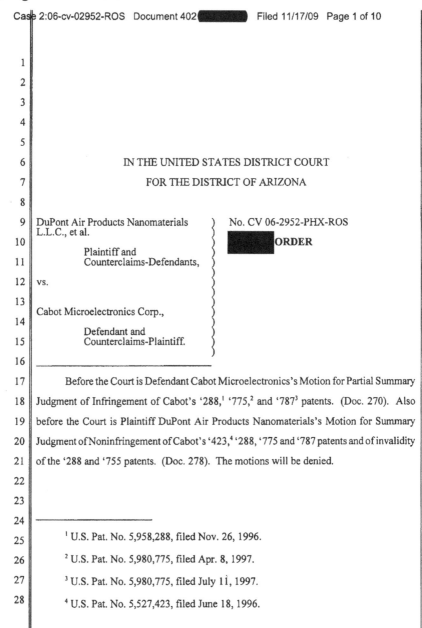

1

2

3

4

5

6   IN THE UNITED STATES DISTRICT COURT

7   FOR THE DISTRICT OF ARIZONA

8

9   DuPont Air Products Nanomaterials   )   No. CV 06-2952-PHX-ROS
    L.L.C., et al.
10   )   ▮▮▮▮ORDER
            Plaintiff and   )
11   Counterclaims-Defendants,   )

12   vs.   )

13   )
    Cabot Microelectronics Corp.,   )
14   )
            Defendant and   )
15   Counterclaims-Plaintiff.   )
)
16   _____

17   Before the Court is Defendant Cabot Microelectronics's Motion for Partial Summary

18   Judgment of Infringement of Cabot's '288,[1] '775,[2] and '787[3] patents. (Doc. 270). Also

19   before the Court is Plaintiff DuPont Air Products Nanomaterials's Motion for Summary

20   Judgment of Noninfringement of Cabot's '423,[4] '288, '775 and '787 patents and of invalidity

21   of the '288 and '755 patents. (Doc. 278). The motions will be denied.

22

23

24   _____

25   [1] U.S. Pat. No. 5,958,288, filed Nov. 26, 1996.

26   [2] U.S. Pat. No. 5,980,775, filed Apr. 8, 1997.

27   [3] U.S. Pat. No. 5,980,775, filed July 1i, 1997.

28   [4] U.S. Pat. No. 5,527,423, filed June 18, 1996.

1   **BACKGROUND**

2   Plaintiff Dupont Air Products Nanomaterials ("DAPN") and Defendant Cabot

3   Microelectronics Corp. ("CMC") market competing slurries for use in chemical-mechanical

4   polishing ("CMP"). (Doc. 279 at 1).

5   CMP is used in integrated circuit ("IC") manufacturing. (Doc. 271 ¶ 4). ICs are

6   manufactured in layers, and CMP is used to make new layers flat so that further layers may

7   be added on top of them. (*See id.* at ¶¶ 5-9). As the name "chemical mechanical polishing"

8   suggests, CMP uses both chemical reaction and mechanical polishing to perform this

9   function. (*Id.* at 9). Typically the chemical reaction is performed by an oxidizer, and the

10 mechanical polishing is done by fine abrasive particles, with the two being combined in an

11 aqueous slurry. (*Id.*).

12   CMC holds the four patents at issue in this suit, which concern improvements to CMP

13 slurries. The '423 patent teaches the use of extremely fine abrasive particles made of

14 alumina to create a stable colloid in which the alumina particles do not clump or precipitate

15 out. (Doc. 297 ¶¶ 500-502). The '288, '775 and '787 patents teach the use of a catalyst with

16 multiple oxidative states, in three different manners, to synergistically enhance the

17 performance of CMP slurries. (Doc. 271 ¶¶ 16-25).

18   DAPN brought suit seeking a declaratory judgment that neither DAPN's Microplanar

19 CMP 3600 ("3600") product nor DAPN's Microplanar CMP 3700M ("3700") product

20 infringe CMC's '142,[5] '288, '785, or '787 patents. DAPN also argues that some of those

21 patents are invalid or unenforceable. (Doc. 1 at 3-11). CMC then filed counter-claims

22 alleging DAPN's 3600 and 3700 products willfully infringe the '288, '775 and '787 patents,

23

24

25

26

_____

27   [5] U.S. Patent No. 4,954,142. DAPN and CMC contested infringement of this patent.
28 (Doc. 23 at 13). Claims related to the '142 are not presently before the Court.

                                      - 2 -

1   and that DAPN's MicroPlanar 3200 series CMP products infringe CMC's '423 patent.[6]

2   (Doc. 73 at 10-12).

3   **ANALYSIS**

4   **I. Jurisdiction**

5   DAPN's complaint alleges that prior to this suit, CMC informed DAPN that the

6   3600 and 3700 products infringed CMC's patents, demanded that DAPN cease selling

7   those products, and intimated imminent litigation if DAPN did not. (Doc. 1 at 2-3).

8   Therefore, it is undisputed that the "adverse legal interests, of sufficient immediacy and

9   reality to warrant the issuance of a declaratory judgment" required to satisfy Article III's

10   case-or-controversy are present here. *MedImmune, Inc. v. Genentech, Inc.*, 549 U.S. 118,

11   127 (2007). Subject matter over actions for a declaratory judgment of patent non-

12   infringement comes from 28 U.S.C. § 1338 (conferring federal jurisdiction over patent

13   issues).

14   **II. Summary Judgment Standard**

15   A court must grant summary judgment if the pleadings and supporting documents,

16   viewed in the light most favorable to the non-moving party, "show that there is no

17   genuine issue as to any material fact and that the moving party is entitled to a judgment as

18   a matter of law." Fed.R.Civ.P. 56(c). Substantive law determines which facts are

19   material, and "[o]nly disputes over facts that might affect the outcome of the suit under

20   the governing law will properly preclude the entry of summary judgment." *Anderson v.*

21   *Liberty Lobby, Inc.*, 477 U.S. 242, 248 (1986). In addition, the dispute must be genuine;

22

23

---

24   [6] CMC also filed a third-party complaint against Precision Colloids. (Doc. 23 at 11-

25   13). CMC later amended its answer and counterclaims to add The Virkler Company as a
defendant, and to join Precision Colloids as a co-defendant instead of as a third-party

26   defendant. (Doc. 72). Precision Colloids is alleged to provide components for DAPN's 3600

27   and 3700 products, and Virkler is alleged to provide components for DAPN's 3200 products.
(Doc. 73 at 10). Per DAPN's motion, the three counterclaim-defendants are collectively

28   referred to as "DAPN." (Doc. 278 at 1).

Case 2:06-cv-02952-ROS   Document 402 ▮▮▮▮▮   Filed 11/17/09   Page 4 of 10

1   that is, "the evidence is such that a reasonable jury could return a verdict for the

2   nonmoving party." *Id.*

3   The affidavit of an expert can be used to defeat summary judgment. *Rebel Oil Co.,*

4   *Inc. v. Atl. Richfield Co.*, 51 F.3d 1421, 1435 (9th Cir. 1995). However, the expert must

5   be competent, and while the underlying factual details and reasoning need not be

6   disclosed, the affidavit must provide sufficient factual basis for the opinion. *Id.* An

7   expert opinion cannot defeat summary judgment if it would not support a favorable jury

8   verdict. *Id.*

9   **III. The '288 and '775 Patents Have Not Been Shown To Be Invalid**

10   Before addressing whether certain patents have been infringed, the Court must first

11   determine the validity of the patents at issue. DAPN claims that both the '288 and '775

12   patents are invalid. A patent is presumed valid at every stage of litigation. *Abbott Labs.*

13   *v. Sandoz, Inc.*, 544 F.3d 1341, 1346 (Fed. Cir. 2008). The invalidity of a patent is a

14   question of fact, and summary judgment of invalidity is proper if there is no disputed

15   material fact and "no reasonable jury could find the patent is not anticipated." *Zenith*

16   *Elecs. Corp. v. PDI Comm. Sys., Inc.*, 522 F.3d 1348, 1356-57 (Fed. Cir. 2008).

17   **A. DAPN Has Not Proven The '288 Patent Is Invalid**

18   DAPN argues that U.S. Patent No. 5,695,384 ("the Beratan patent") anticipates the

19   asserted claims of the '288 patent, thereby rendering the patent invalid. (Doc. 278 at 10-

20   18). The parties do not dispute that the Beratan patent is prior art under 35 U.S.C.

21   § 102(e).[7] (*Id.* at 11). For prior art to invalidate a claim by anticipating it, the prior art

22   must "describe every element of the claimed invention, either expressly or inherently,

23   such that a person of ordinary skill in the art could practice the invention without undue

24   experimentation." *Advanced Display Sys., Inc. v. Kent State Univ.*, 212 F.3d 1272, 1282

25   (Fed. Cir. 2000). The Beratan patent discloses and claims a great number of possible

26   _____

27   [7] According to 35 U.S.C. § 102(e), once a patent application is published or results in
an issued patent, the subject matter disclosed in the application becomes prior art as of the

28   effective filing date of the application.

1   slurries containing at least one salt.[8] (Doc. 297 ¶¶ 402-05). However, it does not claim or

2   disclose any of them as being a catalyst, much less a "catalyst having multiple oxidative

3   states," which is an element of all the claims in the '288 patent. In fact, some of the salts

4   disclosed have only one oxidative state. (Doc. 310 at 18). Nor does the Beratan patent

5   disclose a slurry containing iron salts as species: it only suggests that "[c]ompounds of

6   halides [chlorides, fluorides, iodides or bromides] with sodium, potassium, iron and

7   ammonia, as well as other metals, *may* be suitable" in a CMP slurry. Patent No.

8   5,695,384, col. 4, lines 6-9. Broadly put, the Beratan patent discloses a genus of slurries

9   and some species within that genus.[9]

10   A patent covering a broad genus is "an invitation to find the beneficial species in

11   [that] genus." *Integra Lifesciences I, Ltd. v. Merck KgaA*, 496 F.3d 1334, 1352 (Fed. Cir.

12   2007). So a genus claim does not necessarily bar a patent on a claim to a later-discovered

13   beneficial species: "One of the simplest, clearest, soundest, and most essential principles

14   of patent law, is that a later invention may be validly patented, altho dominated[10] by an

15   earlier patent." *Application of Sarret*, 327 F.2d 1005, 1014 (C.C.P.A. 1964) (*quoting*

16   Emerson Stringham, *Double Patenting*). Accordingly, a disclosed genus cannot

17   anticipate a species within it unless the genus is small. *Atofina v. Great Lakes Chem.*

18   *Corp.*, 441 F.3d 991, 999 (Fed. Cir. 2006). The Beratan genus is not small. (Doc. 297 ¶¶

19   402-05).

20   The '288 patent claims just such beneficial species and it is not invalid as

21   anticipated by the genus claimed and disclosed in Beratan. Additionally, DAPN does not

22

23   ██

24   ██

25   [9] A genus is a grouping of related things, and a species is a subset of the things that
26   are included in a genus.

27   [10] A later patent is dominated by a prior patent when any process or object that
     infringes the later patent must necessarily infringe on the prior patent. Thus, a patentee may
28   be entitled to a patent, but be unable to practice the patent absent rights to use a prior patent.

1   allege that the species disclosed in the Beratan patent have all of the limitations of the

2   claims in the '288 patent and, therefore, does not allege that those species anticipate the

3   '288 claims. *Titanium Metals Corp. v. Banner*, 778 F.2d 775, 780-81 (Fed. Cir. 1985)

4   (stating anticipation requires claims "read on or encompass" that "which was already

5   known"). Finally, nothing in the Beratan patent claims or discloses the salts as a catalyst

6   with multiple oxidative states.

7   　　In summary, DAPN has not shown that the Beratan disclosure would have enabled

8   a person possessing the ordinary skill in the art at that time to practice the species claimed

9   in the '288 patent without undue experimentation. Accordingly, the Beratan patent does

10 not invalidate the '288 patent by anticipation.

11 　　**B. DAPN Has Not Proven The '775 Patent Is Invalid**

12 　　DAPN also argues that the Beratan patent anticipates the asserted claims of the

13 '775 patent. (Doc. 278 at 19-20). However, all the claims of the '775 patent contain a

14 limitation of "at least one catalyst having multiple oxidative states." As shown in the

15 analysis of the '288 patent's validity, the Beratan patent does not claim or disclose such a

16 catalyst, and accordingly does not invalidate the '775 patent by anticipation.

17 　　DAPN next argues that the Beratan patent makes claims 2, 3 and 19 of the '775

18 patent obvious. However, DAPN's argument suffers from two flaws. First, DAPN fails

19 to disclose any prior art that discloses or suggest the "catalyst having multiple oxidative

20 states" limitation present in the claims of the '775 patent but absent in the Beratan patent.

21 And second, a claim is only obvious if "the subject matter as a whole would have been

22 obvious." 35 U.S.C. § 103. It is inappropriate to dissect a claim into discrete elements:

23 the claim as a whole must be considered. *Diamond v. Diehr*, 450 U.S. 175, 188-89

24 (1981). Further, "[t]he question . . . is not whether the differences themselves would

25 have been obvious. Consideration of differences . . . is but an aid in reaching the ultimate

26 determination of whether the claimed invention *as a whole* would have been obvious."

27 *Stratoflex, Inc. v. Aeroquip Corp.*, 713 F.2d 1530, 1537 (Fed. Cir. 1983) (emphasis

28 added). DAPN only argues that the individual elements of the disputed claims in the '775

- 6 -

1   patent would have been obvious to one skilled in the art; no analysis of each claim *as a*

2   *whole* is presented. (Doc. 278 at 19-20; Doc. 348 at 11). Because DAPN's obviousness

3   argument does not address each claim as a whole, DAPN's argument is insufficient as a

4   matter of law to demonstrate that the claims in the '775 patent are obvious. Accordingly,

5   DAPN has not shown that the '775 patent is invalid.

6   **IV. Noninfringement Of The '423 Patent Has Not Been Demonstrated**

7   DAPN seeks summary judgment that its products do not infringe the '423 patent.

8   The '423 patent claims CMP slurries containing alumina particles, and methods of using

9   CMP slurries containing alumina particles. All of the claims contain a limitation that the

10   slurries contain alumina particles have "an aggregate size distribution less than about 1.0

11   micron." The parties dispute the meaning of this limitation, ▮▮▮▮▮▮▮▮▮▮▮

12   ▮▮▮▮▮▮▮▮▮▮▮▮▮▮▮▮▮▮▮▮▮▮▮▮▮▮▮▮▮▮▮

13   As a threshold matter, the transition phrase of the '423 claims must be considered.

14   The transition phrase is critical to interpretation of a claim. "'Consisting of' is a term of

15   patent convention meaning that the claimed invention contains only what is expressly set

16   forth in the claim." *Norian Corp. v. Stryker Corp.*, 363 F.3d 1321, 1331 (Fed. Cir. 2004).

17   But when, as here, the transition phrase is "comprising,"[11] infringement requires only that

18   all the elements of the claim be present, *regardless* of whether additional elements are

19   also present. *Genentech, Inc. v. Chiron Corp.*, 112 F.3d 495, 501 (Fed. Cir. 1997).

20   DAPN argues that it is entitled to a judgment of non-infringement solely because

21   ▮▮▮▮▮▮▮▮▮▮▮▮▮▮▮▮▮▮▮▮▮▮▮▮▮▮▮▮▮▮▮

22   ▮▮▮▮▮ (Doc. 278 at 21-27) But DAPN's argument is without legal merit: to

23   show noninfringement, DAPN must instead show that one or more of the elements in

24   ─────────────────────────

25   [11] "A chemical-mechanical polishing slurry for polishing a metal layer *comprising*:
high purity, alumina particles uniformly dispersed in an aqueous medium having a surface
26   area ranging from about 40 m^2/g to 430 m^2/g, an aggregate size distribution less than about
27   1.0 micron, . . . wherein said slurry is colloidally stable. U.S. Patent No. 5,527,423, claim
17 (filed Oct. 6 1994) (emphasis added). Nothing in the claim requires that the slurry not
28   have alumina particles larger than the claimed range.

1   each of the asserted claims are *not* present; whether additional elements are present is

2   irrelevant. *Mannesmann Demag Corp. v. Engineered Metal Products Co., Inc.*, 793 F.2d

3   1279, 1282-83 (Fed. Cir. 1986). ████████████████████████

4 ██

5 ██

6 ████████ Therefore, DAPN's assertion ████████████████████

7 ████████ does not demonstrate that those products do not infringe the claims of the '423

8   patent.

9   **V. Judgment Regarding Infringement of the '288, '775, and '787 Patents Cannot Be**

10   **Granted**

11   The parties' motions for summary judgment regarding the '288, '775, and '787

12   patents turn on a limitation common to the claims of all three patents: the presence of at

13   least one "catalyst having multiple oxidative states." The issue is whether DAPN's 3600

14   and 3700 products contain such a catalyst and, therefore, infringe the patents.

15   DAPN first alleges that the specification "catalyst having multiple oxidative

16   states" should be limited to a "catalyst having multiple oxidative states *while being used*

17   *for CMP*," (Doc. 278 at 11) or a "catalyst having at least one oxidative state that is higher

18   than the default state." (*Id.* at 5-6 (paraphrased).) Neither argument is correct. Those

19   limitations are not present in the claims, and limitations not in the claims may not be

20   imported from the specification. *CollegeNet, Inc. v. ApplyYourself, Inc.*, 418 F.3d 1225,

21   1231-32 (Fed. Cir. 2005). Therefore, the "catalyst having multiple oxidative states"

22   limitation reads on any catalyst that can have more than one oxidative state without, as

23   required by the Court's construction of the claims, being consumed or undergoing a

24   chemical change."[12] (Doc. 223 at 12) For example, if—as DAPN claims—the default

25

26   [12] In the claims' context of "a catalyst having multiple oxidative states," clearly a

27 change in oxidative state alone cannot be considered consumption or chemical change of the
catalyst, for to do so would render the claims a nullity, in controversion of the presumed

28   validity of the claims.

1 oxidative state of its catalyst is +3, DAPN's catalyst would fall within the limitations of

2 the claims if the catalyst could be coerced into the +2 or +4 oxidative state without being

3 consumed or chemically changed.

4   The parties dispute whether DAPN's catalyst has multiple oxidative states. ███

5 ██

6 ██

7 ████████████████████████   In fact, it appears that DAPN admitted to the patent office

8 that its catalyst has multiple oxidative states. ████████████████████████████████

9 ██

10 ██

11 ████████████████████████████████   Despite the Court's serious misgivings about

12 DAPN's evidence, summary judgment is not the time for the weighing of this evidence.

13 The issue of multiple oxidative states must be resolved by a fact finder.[13]

14   Accordingly,

15   **IT IS ORDERED** Defendant Cabot Microelectronics's Motion for Partial

16 Summary Judgment (Doc. 270) is **DENIED.**

17   **IT IS FURTHER ORDERED** that Plaintiff DuPont Air Products Nanomaterials's

18 Motion for Summary Judgment (Doc. 278) is **DENIED.**

19   **IT IS FURTHER ORDERED** the Joint Motion to Seal (Doc. 289) is **GRANTED**

20   **IT IS FURTHER ORDERED** the Motion for Substitution of Counsel (Doc. 361)

21 is **DENIED AS MOOT.**

22   **IT IS FURTHER ORDERED** the Motion for Leave to File Reply and Motion to

23 Seal (Doc. 396, 397) are **DENIED.**

24 ──────────────────

25   [13] ██

      ██ But

26 DAPN credibly disputes ███

27 ██

28 ████████ There is a material issue of fact on this argument as well.

- 9 -

1　　　　**IT IS FURTHER ORDERED** the parties shall submit their pretrial filings no

2　later than December 16, 2009.

3　　　　**IT IS FURTHER ORDERED** the parties shall submit a joint statement of no

4　more than two pages setting forth whether the Court need resolve the discovery disputes

5　previously identified by the parties. This statement shall be filed no later than November

6　23, 2009.

7

8　　　　DATED this 16th day of November, 2009.

9

10

11

12

13

14

15

16　　　　　　　　　　　　　　　　Roslyn O. Silver
　　　　　　　　　　　　　　　　United States District Judge

17

18

19

20

21

22

23

24

25

26

27

28

- 10 -

IN THE UNITED STATES DISTRICT COURT
FOR THE NORTHERN DISTRICT OF ILLINOIS
EASTERN DIVISION

DSM DESOTECH INC.,	)	
	)	
Plaintiff,	)	
	)	**No. 08 CV 1531**
v.	)	
	)	**Judge Joan H. Lefkow**
3D SYSTEMS CORPORATION and	)	
3D SYSTEMS, INC.,	)	
	)	
Defendants.	)	
	)	

MEMORANDUM OPINION AND ORDER

This case arises out of an eight-count complaint filed by plaintiff, DSM Desotech, Inc. ("Desotech"), against defendants, 3D Systems Corporation and 3D Systems, Inc. (collectively, "3DS"), for violations of federal antitrust law, state antitrust law, state tort law, and federal patent law. Before the court is 3DS's motion to dismiss Counts I through VII of Desotech's complaint, the antitrust and state law claims, pursuant to Rule 12(b)(6) of the Federal Rules of Civil Procedure. For the reasons set forth below, 3DS's motion to dismiss [#29] will be granted in part and denied in part.

BACKGROUND

3DS is a manufacturer of large-frame stereolithography ("SL") machines. SL is a process by which a physical object, such as a model, is created layer by layer from liquid resin that is solidified into shape with a laser. Because the end product created by the SL process is dependent in large part on the quality and specifications of the resins used to create it, a substantial amount of research and development is devoted to the creation of new types of resins. Desotech is a leader in the SL resin market and the holder of two equipment patents allegedly covering the resin recoating technology used in eight of the SL machines produced and sold by 3DS. 3DS likewise produces resins that can be used in the machines it sells.

Desotech alleges that since 2007, 3DS has engaged in unlawful tying, in violation of section 1 of the Sherman Act, 15 U.S.C. § 1, and section 3 of the Clayton Act, 15 U.S.C. § 14, (Counts I and II) by conditioning the sale and maintenance of its large-frame SL machines on the purchase of 3DS's resins. According to the complaint, 3DS "has expressly refused to sell and service its most recent large-frame SL machine—the Viper Pro SLA System—unless customers exclusively purchase resin from 3DS." Am. Comp. ¶ 5.

According to Desotech, 3DS has begun informing customers that only "licensed" or "approved" resins can be used in the Viper Pro, though 3DS neither told customers about the resin limitation at the time of sale nor informed competing resin suppliers about the requisite approval process. To enforce this mandate, 3DS has included in the Viper Pro a radio frequency identification, or "RFID," feature which, if activated by 3DS, will prevent the machine from working should customers attempt to use competing brands of resin. Although customers were aware that the RFID component existed, they allegedly did not know this feature could be used to preclude the use of competing resins that they may want to use. Furthermore, 3DS did not attempt to activate the RFID feature until recent software updates were made to machines already purchased. Regarding those Viper Pro machines in which 3DS has not already activated the RFID feature, 3DS has told its customers that it soon intends to do so. 3DS has warned customers that if they refuse to allow the RFID feature to be activated or if they continue to use unapproved resin, the warranty supplied by 3DS at the time of purchase will be voided.

2

Desotech further alleges that although 3DS used to make several other large-frame SL machines that do not have an RFID component, 3DS has stopped manufacturing those models and, moreover, is attempting to systematically eliminate them from the market. For example, 3DS has allegedly reached an agreement with a leading SL machinery maintenance contractor, National RP Support, Inc. ("National RP"), to stop servicing those older large-frame machines for which no contractual servicing obligation exists. Desotech also asserts that 3DS is removing existing large-frame SL machines from the market by offering substantial incentives to customers to trade-in old machines and purchase a Viper Pro.

Desotech alleges that in addition to unlawful tying, 3DS's contracting and licensing practices amount to an unlawful restraint of trade and attempted monopolization of the SL resin market in violation of sections 1 and 2 of the Sherman Act, 15 U.S.C. §§ 1 and 2 (Counts III and IV, respectively), and the Illinois Antitrust Act, 740 Ill. Comp. Stat. 10/3 (Count V).

Desotech's complaint further alleges that 3DS has made false, misleading, and disparaging statements to Desotech customers about the quality and fitness for use of Desotech's resins, in violation of the Illinois Uniform Deceptive Trade Practices Act, 815 Ill. Comp. Stat. 510/2(a)(7)–(8), and that 3DS has tortiously interfered with prospective economic advantages reasonably anticipated by Desotech, in violation of Illinois state law.

3DS moves to dismiss all of the antitrust and state law claims set forth in the complaint.

STANDARD OF REVIEW

Defendants bring their motion under Rule 12(b)(6) for failure to state a claim upon which relief may be granted. The Supreme Court recently addressed the proper application of the federal notice pleading standards, particularly in regard to antitrust actions, in *Bell Atlantic*

3

Corp. v. *Twombly*, 550 U.S. 544, 127 S. Ct. 1955, 167 L. Ed. 2d 929 (2007). In *Twombly*, the

Court "retire[d]" the frequently quoted language of *Conley* v. *Gibson*, 355 U.S. 41, 45–46, 78 S.

Ct. 99, 2 L. Ed. 2d 80 (1957), "that a complaint should not be dismissed for failure to state a

claim unless it appears beyond doubt that the plaintiff can prove no set of facts in support of his

claim which would entitle him to relief." *Twombly*, 127 S. Ct. at 1968–69 ("The phrase is best

forgotten as an incomplete, negative gloss on an accepted pleading standard: once a claim has

been stated adequately, it may be supported by showing any set of facts consistent with the

allegations in the complaint."). Instead, at the pleading stage, there must be "allegations

plausibly suggesting" an antitrust violation; the mere possibility of later "unearthing direct

evidence" through discovery is not enough to preclude dismissal. *Twombly*, 127 S. Ct. at 1966,

1968.

The Seventh Circuit has explained that despite "some language that could be read to

suggest otherwise, the Court in *Twombly* made clear that it did not, in fact, supplant the basic

notice-pleading standard." *Tamayo* v. *Blagojevich*, 526 F.3d 1074, 1083 (2008); *see also*

Twombly, 127 S. Ct. at 1973, n.14 (expressly disclaiming the establishment of "any 'heightened'

pleading standard" or broadening of the scope of Fed. R. Civ. P. 9); *Lang* v. *TCF Nat'l Bank*,

249 Fed. Appx. 464, 466–67 (7th Cir. 2007) (noting that notice pleading remains the pleading

standard). A plaintiff still must provide only "enough detail to give the defendant fair notice of

what the claim is and the grounds upon which it rests, and, through his allegations, show that it is

plausible, rather than merely speculative, that he is entitled to relief." *Lang*, 249 Fed. Appx. at

466 (internal quotation marks and ellipses omitted) (citing *EEOC* v. *Concentra Health Care*

Servs., Inc., 496 F.3d 773, 776–77 (7th Cir. 2007) (citing *Twombly*, 127 S. Ct. at 1964)).

4

For complaints involving complex litigation—for example, antitrust or RICO claims—a fuller set of factual allegations may be necessary to show that plaintiff's claims are plausible. *Limestone Dev. Corp.* v. *Village of Lemont, Ill.*, 520 F.3d 797, 803 (7th Cir. 2008). Nevertheless, a plaintiff's complaint need only provide a "short and plain statement of the claim showing that the pleader is entitled to relief," Fed. R. Civ. P. 8(a)(2), sufficient to provide the defendant with "fair notice" of the claim and its basis. *Twombly*, 127 S. Ct. at 1964. In addressing a rule 12(b)(6) motion, the court construes the complaint in the light most favorable to the plaintiff, accepting as true all well-pleaded factual allegations and drawing all reasonable inferences in her favor. *Tamayo*, 526 F.3d at 1081.

DISCUSSION

I. Tying Claims (Counts I and II)

Desotech has filed two counts against 3DS for unlawful tying, the first under § 1 of the Sherman Act (Count I) and the second under § 3 of the Clayton Act (Count II). Although some older cases state otherwise, the standards for adjudicating tying claims under the two statutes are now recognized to be the same. *Sheridan* v. *Marathon Petroleum Co.*, 530 F.3d 590, 592 (7th Cir. 2008). In a tying agreement, a seller conditions the sale of a product or service on the buyer's purchase of another product or service from, or by direction of, the seller. *Id.* at 592. Of course, every refusal to sell two products separately cannot be said to restrain competition. *Reifert* v. *S. Cent. Wis. MLS Corp.*, 450 F.3d 312, 322 (7th Cir. 2006). If each of the products may be purchased separately in a competitive market, one seller's decision to sell the two in a single package imposes no unreasonable restraint on either market. *D.O. McComb & Sons, Inc.* v. *Memory Gardens Mgmt. Corp., Inc.*, 736 F. Supp. 952, 957 (N.D. Ind. 1990).

5

The Supreme Court has established that "the essential characteristic of an invalid tying arrangement lies in the seller's exploitation of its control over the tying product to force the buyer into the purchase of a tied product that the buyer either did not want at all, or would have preferred to purchase elsewhere on different terms." *Jefferson Parish Hosp. Dist. No. 2* v. *Hyde*, 466 U.S. 2, 12, 104 S. Ct. 1551, 80 L. Ed. 2d 2 (1984), *abrogated on other grounds by Ill. Tool Works, Inc.* v. *Indep. Ink, Inc.*, 547 U.S. 28, 126 S. Ct. 1281, 164 L. Ed. 2d 26 (2006). Where such "forcing" is present, "competition on the merits in the market for the tied item is restrained and the Sherman Act is violated." *Id.* In order to establish the per se illegality of a tying arrangement, a plaintiff must demonstrate that (1) the tying arrangement is between two distinct products or services, (2) the defendant has sufficient economic power in the tying market to appreciably restrain free competition in the market for the tied product, and (3) a not insubstantial amount of interstate commerce is affected. *Reifert*, 450 F.3d at 316 (7th Cir. 2006). In addition, the Seventh Circuit has held that "an illegal tying arrangement will not be found where the alleged tying company has absolutely no economic interest in the sales of the tied seller, whose products are favored by the tie-in." *Carl Sandburg Vill. Condo. Ass'n No. 1* v. *First Condo. Dev. Co.*, 758 F.2d 203, 207–08 (7th Cir. 1985); *accord Reifert*, 758 F.2d at 316.

Defendants contend that Desotech's complaint fails to adequately allege the first three elements of an unlawful tying arrangement. Because the court agrees that Desotech did not adequately plead the third element—that the tying arrangement forecloses a substantial volume of commerce—the other elements need not be addressed at this time.

The ultimate flaw in Desotech's tying claim is its failure to allege the existence of a coerced tying arrangement on the part of 3DS. As the Supreme Court has explained, "the

6

essential characteristic of an invalid tying arrangement lies in the seller's exploitation of its control over the tying product to *force* the buyer into the purchase of a tied product" he does not want. *Jefferson Parish*, 466 U.S. at 12 (emphasis added). If the consumer remains free to buy the tying product without also buying the product to which it is ostensibly tied, then no coercion has occurred. *See id.* at 12 n.6 (citing *N. Pac. Ry. Co.* v. *United States*, 356 U.S. 1, 6 n. 4, 78 S. Ct. 514, 2 L. Ed. 2d 545 (1958)). Even if the defendant attempts to induce consumers into purchasing both products he sells by offering the bundle at discount prices, he has not engaged in illegal tying so long as the consumer remains free to purchase each item separately. *See Schor* v. *Abbott Labs.*, 457 F.3d 608, 610 (7th Cir. 2006) (drug manufacturer did not effect a tying arrangement when selling a patented drug in combination with another drug at a discount because patented drug was also available for sale separately). This is because antitrust law is only concerned with behavior that has "a substantial potential for impact on competition." *Jefferson Parish*, 466 U.S. at 16. Furthermore, "[i]f only a single purchaser were 'forced' with respect to the purchase of a tied item, the resultant impact on competition would not be sufficient to warrant the concern of antitrust law." *Id.*

Many circuits thus require plaintiffs alleging tying claims to show not only that a defendant has market power but also that the defendant has wielded such market power to force consumers to alter their purchasing choices. *See, e.g., United States* v. *Microsoft Corp.*, 253 F.3d 34, 85 (D.C. Cir. 2001); *Thompson* v. *Metro. Multi-List, Inc.*, 934 F.2d 1566, 1577 (11th Cir. 1991). Although the Seventh Circuit does not explicitly require evidence of coercion as an independent element of tying claims, Judge Wood has recognized that demonstrating foreclosure of competition in the tied product market—the third element of the Seventh Circuit's test—is essentially equivalent to the "coercion element" required by other circuits. *Reifert*, 450 F.3d at 323 (Wood, J., concurring). After all, if the defendant has not forced consumers into the tying arrangement alleged, it can hardly be said that such arrangements have substantially affected interstate commerce to the detriment of those consumers.

Here, Desotech's complaint fails to adequately allege coercion on the part of 3DS. Although Desotech states that 3DS "expressly refuses to sell and service" its newest line of SL machines unless customers purchase their resin exclusively from 3DS, Am. Compl. ¶ 5, Desotech's allegations regarding particular customers fail to support such a conclusion.

Desotech alleges that 3DS told two Desotech customers, Express Pattern and Dynacept, that they could not use newer Desotech resins in their Viper Pro machines. Desotech further alleges that 3DS told Dynacept that if it continued to use "non-approved" Desotech resins in its Viper Pro machines, 3DS would withhold maintenance service and void existing warranties. Am. Compl. ¶ 63.

Although such actions on the part of 3DS might be considered a breach of warranty (or otherwise anticompetitive, as discussed in further detail below), such allegations do not amount to a tie-in. *See, e.g.*, *Va. Panel Corp.* v. *MAC Panel Co.*, 133 F.3d 860, 870–71 (D.C. Cir. 1998) ("[V]oiding a warranty on a product already sold, while possibly a breach of warranty, cannot be a tying arrangement because the purchaser is not deciding whether to buy a product."). To the extent Desotech complains that 3DS is tying the sale of its machine to its resins, such assertions cannot be supported by 3DS's actions towards Express Pattern and Dynacept, both of which had already purchased their Viper Pros at the time the alleged tie-in was executed. *See N. Pac. Ry.*, 356 U.S. at 5–6 ("For our purposes a tying arrangement may be defined as an agreement by a

8

party to *sell one product but only on the condition* that the buyer also purchases a different (or

tied) product, or at least agrees that he will not purchase that product from any other supplier.")

(emphasis added).

Furthermore, to the extent that 3DS alleges that service or maintenance of the machines

is the tying market at issue for these consumers,[1] Desotech has failed to identify, much less

adequately allege, a market for such service, as would be required in order to allege that 3DS has

market power in the tying market.[2] *Cf.* Am. Compl. ¶ 82 (*"Large-frame SL machines (i.e., the*

tying product) and resins for those machines (i.e., the tied product) constitute separate and

distinct products.") (emphasis added).

Nor are Desotech's tying allegations saved by its pleadings with respect to Tangible

Express or AP Pronto ("APP"), two other Desotech customers. Desotech alleges that as part of a

transaction in which Tangible Express purchased six Viper Pro machines from 3DS, those

companies executed a memorandum of understanding that "expressly required Tangible Express

to purchase all of its resin for the six Viper Pro machines directly from 3DS."[3] Am. Compl. ¶

68. While this allegation clearly alleges a contract of exclusive dealing between 3DS and

Tangible Express, Desotech does not allege that 3DS required such exclusive dealing as a

---

[1] Although the complaint never explicitly identifies maintenance of large-frame SL machines as a relevant "tying market," it repeatedly alleges that 3DS *"condition[ed] the* sale and/or *maintenance* of its large-frame SL machines on the purchase of its own SL resins." Am. Compl. ¶¶ 80, 90, 94.

[2] The complaint suggests that there are at least a few third-party maintenance contractors who could provide service to Viper Pro owners independent of 3DS. While one such contractor has allegedly agreed with 3DS not to provide service to Viper Pro owners who use Desotech resins, the court is unable to assess a maintenance/resin tying claim absent further factual detail about the market for SL machine service.

[3] Although 3DS submitted along with its motion to dismiss a copy of the memorandum of understanding, the court finds Desotech's allegations inadequate even without consideration of the submitted exhibit and, therefore, will not consider it at this time.

condition on the sale of its Viper Pro machines. Similarly, Desotech alleges that in a recent deal

between 3DS and APP, "3DS required APP to trade-in an older SL machine for a Viper Pro and

to purchase its resins exclusively from 3DS rather than Desotech." Am. Compl. ¶ 70. Desotech

once again failed, however, to allege that such exclusive dealing was a precondition to the

availability of 3DS's new Viper Pro.[4] It could be consistent with lawful behavior on the part of

3DS if it won those exclusive contracts through aggressive selling techniques or discounts on its

products. *See Waldo* v. *N. Am. Van Lines, Inc.*, 669 F. Supp. 722, 731 (W.D. Pa. 1987)

("[S]imply characterizing an exclusive dealing arrangement as a tying arrangement does not

necessarily make it one.") (citations omitted). Indeed, nowhere does Desotech allege that either

of these customers is unhappy with its current resin sourcing arrangements.

The closest Desotech comes to an adequate allegation of coercion is in its assertions

regarding the interaction between 3DS and Lockheed Martin ("Lockheed"), another Desotech

customer. Desotech alleges that in mid-2007, Lockheed Martin informed 3DS that one of its

SLA 500 machines, a predecessor to the new Viper Pro, stopped working.[5] Desotech further

alleges that "[a]fter a significant amount of pressure from 3DS, including 3DS's refusal to sell

the Viper Pro machine to Lockheed unless Lockheed agreed to purchase resins exclusively from

3DS and to stop purchasing resins from Desotech, Lockheed agreed to purchase a Viper Pro to

---

[4] In its response brief, Desotech asserts that the complaint alleges that "AP Proto, Lockheed and Tangible Express were *required* to purchase their resins exclusively from 3DS as a condition to buying their Viper Pro machines." Pl.'s Resp. at 14 (emphasis in original). Desotech did not, however, include such an assertion in its amended complaint. Furthermore, even if that language had been included in the amended complaint, it is not clear that such a "conclusory allegation," without more factual specificity as to each of the three customers, would be sufficient under *Twombly*, 127 S. Ct. at 1966.

[5] According to the complaint, 3DS has stopped manufacturing its older lines of SL machines, including the SLA 500.

replace its SLA 500 machine." Am. Compl. ¶ 66. Desotech also alleges that Lockheed is unhappy about this arrangement and has expressed to Desotech its desire to use Desotech's resins rather than 3DS's.

With these allegations concerning Lockheed, Desotech has given 3DS notice of its claim and identified a particular set of events from which 3DS could begin to prepare a defense. Most importantly, Desotech alleges that Lockheed did not want to purchase 3DS's resins, but that Lockheed was nevertheless required to if it wanted to continue to purchase and use 3DS's SL machines. Such allegations form the cornerstone of a successful tying claim. Still, these particular allegations alone cannot carry Desotech's burden on this motion to dismiss. As the Supreme Court made clear in *Jefferson Parish*, "[i]f only a single purchaser were 'forced' with respect to the purchase of a tied item, the resultant impact on competition would not be sufficient to warrant the concern of antitrust law," 466 U.S. at 16, and, furthermore, Desotech alleges that Lockheed purchased only a single Viper Pro machine.

Because Desotech has failed to sufficiently allege the existence of an illegal tying arrangement, Counts I and II of its complaint will be dismissed without prejudice.

II. Attempted Monopolization (Count IV)

Section 2 of the Sherman Act provides that "[e]very person who shall monopolize, or attempt to monopolize, or combine or conspire with any other person or persons to monopolize any part of the trade or commerce among the several States, or with foreign nations, shall be deemed guilty of a felony . . . ." 15 U.S.C. § 2. Desotech alleges that 3DS has attempted to monopolize the SL resin market. To prove attempted monopolization, a plaintiff must show "(1) that the defendant has engaged in predatory or anticompetitive conduct with (2) a specific

11

intent to monopolize and (3) a dangerous probability of achieving monopoly power." *Spectrum*

Sports, Inc. v. *McQuillan*, 506 U.S. 447, 456, 113 S. Ct. 884, 122 L. Ed. 2d 247 (1993); *accord*

Indiana Grocery, Inc. v. *Super Valu Stores, Inc.*, 864 F.2d 1409, 1412 (7th Cir. 1989); *Nat'l*

Black Expo v. *Clear Channel Broad., Inc.*, 2007 WL 495307, at *8 (N.D. Ill. Feb. 8, 2007).

A. Dangerous Probability of Success

The "dangerous probability" element requires allegations that 3DS had sufficient market

power to threaten actual monopolization within the relevant market. *Indiana Grocery*, 864 F.2d

at 1413. As discussed below, Desotech's complaint sufficiently alleges that 3DS poses a

dangerous probability of acquiring monopoly power in the SL resin market .

1. Market Power

Desotech alleges that 3DS has the market power to threaten actual monopolization.

Specifically, Desotech alleges that 3DS has a "present and growing market share of over 50%"

of the SL resin market, making it the single largest supplier. Am. Compl. ¶ 9. Although 3DS

cites *Blue Cross & Blue Shield United of Wisconsin* v. *Marshfield Clinic*, 65 F.3d 1406, 1411

(7th Cir. 1995), for the proposition that "[f]ifty percent is below any accepted benchmark for

inferring monopoly power from market share," 3DS has conflated the terms *monopoly* power and

market power. Although the terms are synonymous in most respects and are sometimes used

interchangeably, *see, e.g., Cost Management Services, Inc.* v. *Washington Natural Gas Co.*,

99 F.3d 937, 950 n.15 (9th Cir. 1996), "monopoly power" generally denotes some higher

threshold of market power than does the term "market power" alone.[6] *See, e.g., Eastman Kodak*

---

[6] Whereas the Seventh Circuit has long been comfortable defining monopoly power in simple terms—e.g., "power over price" or "the ability to cut back the market's total output and so raise prices," *Indiana Grocery*, 864 F.2d at 1414—it commented in a recent opinion that "'market power' is key, but its meaning requires elucidation." *Sheridan* v. *Marathon Petroleum*

12

Co. v. *Image Technical Services, Inc.*, 504 U.S. 451, 481, 112 S. Ct. 2072, 119 L. Ed. 2d 265 (1992) ("Monopoly power under § 2 requires, of course, something greater than market power under § 1."); *Bacchus Indus., Inc.* v. *Arvin Indus., Inc.*, 939 F.2d 887, 894 (10th Cir. 1991) ("Monopoly power is also commonly thought of as substantial market power.").

    With respect to a claim for attempted monopolization, a plaintiff need only show that 3DS currently has *market* power and that such market power will tend to approach monopoly power if the alleged unlawful conduct remains unchecked. *See Star Fuel Marts, LLC* v. *Sam's East, Inc.*, 362 F.3d 639, 648 n.3 (10th Cir. 2004) ("[A]n attempted monopolization claim requires . . . that the defendant have sufficient market power such that there is a 'dangerous probability' that an attempt to achieve monopoly power will succeed.") (internal quotation marks omitted) (citing *Brooke Group Ltd.* v. *Brown & Williamson Tobacco Corp.*, 509 U.S. 209, 251, 113 S. Ct. 2578, 125 L. Ed. 2d 168 (1993)). In *Lektro-Vend*, the court held that the defendant's alleged practices "raise[d] a dangerous propensity for creation of an actual monopoly." *Lektro-Vend Corp.* v. *Vendo Co.*, 403 F. Supp. 527, 534 (C.D. Ill. 1975), *aff'd* 545 F.2d 1050 (7th Cir. 1976), *rev'd on other grounds*, 433 U.S. 623, 97 S. Ct. 2881, 53 L. Ed. 2d 1009 (1977). In reaching this determination, the court emphasized that the defendant maintained a "significant market share," "most probably over 20%." *Id.* The court also observed that "the number of competitors [had] been steadily declining" within the relevant market. *Id.*; *see also Hardy* v. *City Optical Inc.*, 39 F.3d 765, 767 (7th Cir. 1994) (noting in a related context that 30% is "the minimum market share from which the market power required to be shown at the threshold of a tying case can be inferred").

Co. LLC, 530 F.3d 590, 594 (7th Cir. 2008).

Here, Desotech similarly alleges that since 3DS began the alleged unlawful behavior in mid-2007, 3DS's market share has been increasing and, consequently, the shares of Desotech and other resin dealers steadily decreasing. At this stage of the litigation, absent the aid of discovery, Desotech need not allege in exact numbers the percentage increase of 3DS's market share. Furthermore, "the Sherman Act's prohibition against attempted monopolization does not require that the attempt in fact ripen into an actual monopoly. It is the attempt which is the offense." *Lektro-Vend Corp.* v. *Vendo Co.*, 660 F.2d 255, 270 (7th Cir. 1981). As alleged by Desotech, the market for SL resin is highly concentrated, with 95% of the market represented by three suppliers—3DS (50%), Desotech (35%), and Huntsman (10%). If Desotech were forced out of the market, 3DS's market power would likely increase substantially.

3DS maintains that Desotech's allegations of market power are nevertheless inadequate because market share alone is insufficient to indicate market power. Even if 3DS's market share alone is an insufficient indicator of its capacity to control prices and exclude competitors, however, consideration of other factors relevant to the dangerous probability analysis—"the alleged offender's ability to achieve the forbidden result, his intent, and the nature of his overt actions," *Lektro-Vend*, 660 F.2d at 271 (internal citations omitted)—indicate that Desotech has adequately alleged the dangerous probability element of its attempted monopolization claim.

For example, 3DS has allegedly leveraged its monopoly power in the market for large-frame SL machines to exclude competitors from the resin market by outfitting its machines with an RFID feature that will render the machine inoperable should the operator attempt to use unapproved resins. Further allegations indicate that 3DS has attempted to systematically remove from the market (or prevent from being serviced) older SL machines that lack the RFID feature.

14

Such actions would undoubtedly increase 3DS's ability to control prices and exclude competitors in the SL resin market. Although 3DS asserts that it has instituted a licensing and qualification process by which competing resin producers might remain active, Desotech has plausibly alleged that the asserted licensing process is in fact a sham.

2. Relevant Market

The relevant market is defined as "the market area in which the seller operates, and to which the purchasers can practicably turn for supplies." *Nat'l Black Expo*, 2007 WL 495307, at *8 (quoting *Tampa Elec. Co.* v. *Nashville Co.*, 365 U.S. 320, 327, 81 S. Ct. 623, 5 L. Ed. 2d 580 (1961)). The relevant market contains both product and geographical components. *Spectrum Sports*, 506 U.S. at 459. The definition of the relevant market is determined based on "the nature of the commercial entities involved and by the nature of the competition that they face." *United States* v. *Phillipsburg Nat'l Bank & Trust Co.*, 399 U.S. 350, 359, 90 S. Ct. 2035, 26 L. Ed. 2d 658 (1970).

Desotech defines the relevant product market as "resins used in [large-frame SL] machines" and the relevant geographic market as "the United States." Am. Compl. ¶¶ 26, 27. As market definitions are questions of fact ordinarily determined at trial, *L&W/Lindco Prods.* v. *Pure Asphalt Co.*, 979 F. Supp. 632, 638 (N.D. Ill. 1997), Desotech has pled a relevant market sufficient to survive a motion to dismiss. *See Todd* v. *Exxon Corp.*, 275 F.3d 191, 203 (2d Cir. 2001) ("At [the pleading] stage, it is sufficient that plaintiff has alleged specific facts that support a narrow product market in a way that is plausible and bears a rational relation to the methodology courts prescribe to define a market for antitrust purposes.").

B. Predatory or Anticompetitive Conduct

The Supreme Court has characterized this element of the attempted monopolization

offense as "the use of monopoly power 'to foreclose competition, to gain a competitive

advantage, or to destroy a competitor.'" *Eastman Kodak Co.* v. *Image Technical Servs., Inc.,*

504 U.S. 451, 482–83, 112 S. Ct. 2072, 119 L. Ed. 2d 265 (1992) (quoting *United States* v.

Griffith, 334 U.S. 100, 107, 68 S. Ct. 941, 92 L. Ed. 1236 (1948)). Predatory conduct has been

broadly defined as conduct "that has no legitimate business justification other than to destroy or

damage competition." *Great Escape,* 791 F.2d at 541. The Seventh Circuit has stated more

generally, however, that when it comes to liability under § 2 of the Sherman Act, "if conduct is

not objectively anticompetitive the fact that it was motivated by hostility to competitors . . . is

irrelevant." *Olympia Equip. Leasing Co.* v. *W. Union Tel. Co.,* 797 F.2d 370, 379 (7th Cir.

1986) (citing *Ball Mem'l Hosp.* v. *Mut. Hosp. Ins., Inc.,* 784 F.2d 1325, 1338–39 (7th Cir.

1986)).

The conduct alleged by Desotech is plainly anticompetitive and intended to foreclose

competition. 3DS cites Judge Posner's antitrust textbook for the proposition that "promoting

customer satisfaction by restricting the supplies that can be used in a seller's equipment to those

that meet the seller's qualification is procompetitive." Defs.' Memo. at 24 (citing RICHARD A.

POSNER, ANTITRUST LAW 201 (2d ed. 2001)). But the allegations set forth in the complaint do

not suggest that 3DS's restrictions on resins used in Viper Pro machines promote customer

satisfaction. On the contrary, Desotech alleges that several of its customers have identified

mechanical problems when using 3DS resins in their Viper Pros. For example, Lockheed is

allegedly unable to use the Desotech resin that it has identified as ideal for its particular purpose;

16

instead, it has been required to use 3DS resins which it has found to be disappointing and "insufficiently accura[te], with particularly bad 'differential shrink,' which is different dimensional changes based on the thickness of the part." Am. Compl. ¶ 67.

Even if the court were to assume, however, that there are some 3DS consumers who are happier with the RFID-laden Viper Pro than they were with machines absent such a feature, 3DS's other alleged actions also foreclose competition. Desotech has alleged that the RFID licensing and qualification procedure is non-existent, which places 3DS in complete control of competition in the resin market for Viper Pros. Desotech has further alleged that because of this licensing inadequacy, 3DS has reduced the number of resins available to Viper Pro owners from over 40 to just a few.

Additionally, Desotech has alleged that 3DS is systematically preventing older SL machines from remaining on the market and that it is actively preventing current Viper Pro owners from disabling the RFID feature. According to Desotech, 3DS has stopped producing non-RFID machines altogether. While a company has no obligation to continue producing older lines of products it no longer deems profitable, Desotech alleges that 3DS is attempting to eliminate the secondary market for SL machines by removing its older products from commerce. *Cf. United States* v. *United Shoe Mach. Corp.*, 110 F. Supp. 295, 334, 343–44 (1922) (finding that defendant had monopolized the shoe machinery market by, in part, purchasing "second-hand machinery mostly of its own manufacture" for the purpose of "curtail[ing] competition from second-hand shoe machinery"), *aff'd* 347 U.S. 521, 74 S. Ct. 699, 98 L. Ed. 910 (1954) (per curiam). If successful, such a scheme could further strengthen 3DS's control over competition in

17

the SL resin market by eliminating from the market any older SL machines, which lack the RFID feature and thus allow for use of competing resins.

For example, Desotech alleges that maintenance contractor National RP has entered into a contract with 3DS that requires National RP "to cease servicing and refurbishing all SLA 500 machines not currently covered by an existing maintenance contract." Am. Compl. ¶ 71. Such a contract has no obvious business justification other than increasing 3DS's market power in the SL machine and resin markets. National RP has also allegedly agreed with 3DS to stop servicing Viper Pro machines used with "non-approved" Desotech resins. Am. Compl. ¶¶ 63, 72. The procompetitive justifications for such a restriction are not readily apparent. For example, if a 3DS customer would rather void the warranty on its Viper Pro than be subject to restrictions on its choice of resin, what legitimate interest would 3DS have in preventing that customer from independently hiring a third-party maintenance provider such as National RP? Ultimately, Desotech will have the burden of proving that the anticompetitive harm from 3DS's conduct outweighs the procompetitive benefits. For purposes of this motion to dismiss, however, Desotech has sufficiently alleged predatory or anticompetitive conduct.

C. Specific Intent

Because "[a]ll lawful competition aims to defeat and drive out competitors," the "mere intention to exclude competition and to expand one's own business is not sufficient to show a specific intent to monopolize." *Great Escape Inc.* v. *Union City Body Co.*, 791 F.2d 532, 541 (7th Cir. 1986) (internal citations omitted). Rather, the acts from which the court can infer specific intent must be essentially predatory in nature. *Id.* Although the specific intent requirement is consistently listed in the Seventh Circuit as a separate inquiry from the predatory

18

conduct element, the same evidence that is used to prove predatory conduct often establishes specific intent as well. *See, e.g.*, *L&W/Lindco*, 979 F. Supp. at 638–39. As discussed above, Desotech has sufficiently alleged predatory conduct, and specific intent—"an intent to control prices and destroy competition," *id.* at 639—can be inferred from the same allegations. *See Great Escape*, 792 F.2d at 541 ("Specific intent may be inferred from predatory conduct . . . .").

D. Antitrust Injury

3DS's last argument for dismissal of Desotech's attempted monopolization claim is based on its contention that Desotech has failed to allege antitrust injury. To have a right of action under the antitrust laws of the United States, a "plaintiff must prove the existence of '*antitrust injury*, which is to say injury of the type the antitrust laws were intended to prevent and that flows from that which makes defendants' acts unlawful.'" *Atl. Richfield Co.* v. *USA Petroleum Co.*, 495 U.S. 328, 334, 110 S. Ct. 1884, 109 L. Ed. 2d 333 (1990) (emphasis in original) (quoting *Brunswick Corp.* v. *Pueblo Bowl-O-Mat, Inc.*, 429 U.S. 477, 489, 97 S. Ct. 690, 50 L. Ed. 2d 701 (1977)). In other words, it is not enough for a plaintiff to show injury in fact that is causally linked to the illegal conduct; the plaintiff must further show injury that "reflect[s] the anticompetitive effect either of the violation or of anticompetitive acts made possible by the violation." *Brunswick Corp.*, 429 U.S. at 489.

The Seventh Circuit has interpreted this standard as requiring that a plaintiff "show that its loss comes from acts that reduce output or raise prices to consumers" in order to sufficiently plead antitrust injury. *See Stamatakis Indus., Inc.* v. *King*, 965 F.2d 469, 471 (7th Cir. 1992) (quoting *Chi. Prof'l Sports Ltd. Partnership* v. *Nat'l Basketball Ass'n*, 961 F.2d 667, 670 (7th Cir. 1992)); *accord U.S. Gypsum Co.* v. *Ind. Gas Co.*, 350 F.3d 623, 626–27 (7th Cir. 2003). If

19

"a victory for the competitor can confer no benefit, certain or probable, present or future, on consumers . . . a court is entitled to question whether a violation of antitrust law is being charged." *Brunswick Corp.* v. *Riegel Textile Corp.*, 752 F.2d 261, 266 (7th Cir. 1984).

Desotech has plainly alleged antitrust injury. Desotech's complaint states that 3DS's anticompetitive behavior forecloses existing competition, discourages innovation, and "dissuades potential new entrants from entering the large-frame SL resin market." Am. Compl. ¶ 77. It alleges further that as a result of 3DS's actions, resin purchasers "will have a substantially smaller selection of resins from which to make their end products." Am. Compl. ¶ 78. Should Desotech ultimately prove these allegations and prevail in this litigation, Desotech's victory would also be a win for consumers.

For these reasons, the court denies 3DS's motion to dismiss Desotech's claim for attempted monopolization, Count IV of the amended complaint.

III. Unreasonable Restraint of Trade (Count III)

In Count III, its claim for unreasonable restraint of trade under section 1 of the Sherman Act, Desotech alleges that in addition to the tying allegations,

> 3DS's conduct in contracting and licensing with its customers constitutes an unreasonable restraint of trade and commerce, in violation of § 1 of the Sherman Act, 15 U.S.C. § 1. In particular, 3DS's unilateral declaration that all machine warranties are void if a customer does not purchase its resins from 3DS and its failure to service a large frame SL machine until a customer switches resin orders to 3DS constitute an unreasonable restraint of trade.

Am. Compl. ¶ 94.

In its memorandum in support of its motion to dismiss, 3DS does not specifically address the sufficiency of Count III. Although 3DS does argue that the complaint fails to make certain allegations that are "necessary for *all* of Desotech's antitrust claims," Defs.' Mem. at 15

20

(emphasis added), the court, in its analysis of the attempted monopolization claim, has already considered and rejected those general arguments.

"The rule of reason is the accepted standard for testing whether a practice restrains trade in violation of § 1." *Leegin Creative Leather Prods., Inc.* v. *PSKS, Inc.*, — U.S. —, 127 S. Ct. 2705, 2712, 168 L. Ed. 2d 623 (2007). Rule of reason analysis of claims arising under section 1 of the Sherman Act is similar to analysis of monopolization claims under section 2. *United States* v. *Microsoft Corp.*, 253 F.3d 34, 59 (D.C. Cir. 2001). Under either analysis, courts apply a balancing approach, weighing the anticompetitive harm of the conduct against the procompetitive benefit. *Id.* The essence of the rule of reason analysis under section 1 is the requirement that the plaintiff "show that the challenged restraint has an adverse impact on competition in a relevant market." *Dos Santos* v. *Columbus-Cuneo-Cabrini Med. Ctr.*, 684 F.2d 1346, 1352 (7th Cir. 1982). Having found (as discussed above) that Desotech sufficiently alleged a claim for attempted monopolization under section 2, the court finds that the same allegations also support, in large part, its claim under section 1.

Of course, in contrast to section 2, section 1 of the Sherman Act reaches only "unreasonable restraints of trade effected by a 'contract, combination . . . or conspiracy' between separate entities" and "does not reach conduct that is 'wholly unilateral.'" *Copperweld Corp.* v. *Independence Tube Corp.*, 467 U.S. 752, 104 S. Ct. 2731, 81 L. Ed. 2d 628 (1984) (quoting 15 U.S.C. § 1). Although Desotech's section 1 allegations are, by their own terms, premised largely on 3DS's "unilateral declaration," *id.* ¶ 94, Desotech further alleges that "[t]he contracts and licenses between 3DS and its customers constitute concerted action." *Id.* ¶ 95. Thus, Desotech's section 1 allegations are not based entirely on unilateral conduct and do not fail on that count.

21

3DS's motion to dismiss Count III is therefore denied.

IV. Illinois Antitrust Act (Count V)

Both parties agree that courts "shall use the construction of the federal law by the federal

courts as a guide in construing [the Illinois Antitrust Act]" when "the wording [of the Act] is

identical or similar to that of federal antitrust law." 740 Ill. Comp. Stat. 10/11. The language of

740 Ill. Comp. Stat. 10/3, the provision of the Illinois Antitrust Act cited in Desotech's state

antitrust count, is similar, in relevant parts, to the Sherman Act and the Clayton Act. The court

will thus look to federal law interpreting the relevant language of those federal antitrust statutes

for guidance in analyzing Desotech's claims under the Illinois Antitrust Act. *A & A Disposal &*

Recycling, Inc. v. *Browning-Ferris Indus. of Ill., Inc.*, 664 N.E.2d 351, 352–53, 279 Ill. App. 3d

337, 215 Ill. Dec. 954 (Ill. App. Ct. 2d Dist. 1996). Because Desotech has stated valid claims for

attempted monopolization under section 2 of the Sherman Act and for unreasonable restraint of

trade under section 1 of the Sherman Act, Desotech has likewise stated a valid claim under 740

Ill. Comp. Stat. 10/3. 3DS's motion to dismiss Count V is therefore denied.

V. Illinois Uniform Deceptive Trade Practices Act (Count VI)

In Count VI, Desotech brings claims for commercial disparagement under sections

2(a)(7) and 2(a)(8) of the Illinois Uniform Deceptive Trade Practices Act (IUDTPA), 815 Ill.

Comp. Stat. 510/2. To state a claim for commercial disparagement under section 2(a)(8), a

plaintiff must allege that the defendant made specific statements "disparag[ing] the goods,

services, or business of another by false or misleading representation of fact." 815 Ill. Comp.

Stat. 510/2(a)(8); *accord Stenograph Corp.* v. *Microcat Corp.*, No. 86 C 10231, 1989 WL

99543, at *4 (N.D. Ill. Aug. 22, 1989). Similarly, under section 2(a)(7), a plaintiff must allege

22

that the defendant "represent[ed] that goods or services are of a particular standard, quality, or grade or that goods are a particular style or model, if they are of another." 815 Ill. Comp. Stat. 510/2(a)(7).

The complaint alleges that 3DS made various disparaging statements and misrepresentations to Desotech customers, including (1) that Desotech "had not done its 'due diligence'" with respect to certain of its resins, Am. Compl. ¶ 54; (2) that those resins "were of insufficient quality to run on 3DS's large-frame SL machines" (though, Desotech further alleges, "3DS itself had previously distributed some of the exact same resins under its distribution agreement with Desotech"), *id.*; (3) that certain Desotech resins were not "licensed" or "qualified" for use in the Viper Pro, though such a licensing or qualification process apparently did not exist, *id.* ¶¶ 55, 56; and (4) that using Desotech resins in the Viper Pro would "cause the machine's software to 'time bomb out' and make the machine inoperable," *id.* ¶ 61.

3DS contends that Desotech's allegation are "conclusory generalizations" that fail to specify any particular statements made by 3DS and, furthermore, are insufficient as a matter of law to show disparagement. 3DS argues, for example, that (1) the alleged statements by 3DS "neither '*criticize the quality* of one's goods or services,' nor describe the attacked service or product as 'substandard, negligent, or harmful,'" but "merely describe performance characteristics (and limitations) of the Viper Pro"; (2) Desotech "fails to show any '*false or misleading* representation of fact' regarding the Viper Pro or the use of newer [Desotech] resins"; and (3) Desotech "fails to allege that [3DS] incorrectly represented that the newer [Desotech] resins 'are of a particular standard, quality, or grade,' as it must to state a claim under [815 Ill. Comp. Stat. 510/2(a)(7)]." Defs.' Mem. at 29–30 (internal citations omitted) (emphasis

23

in original). Construing the allegations in the light most favorable to the plaintiff, as the court

must on a motion to dismiss, the court is not persuaded by any of these arguments; at best, they

raise issues of fact that are more appropriately addressed at later stages of litigation. The court

finds that Desotech's allegations under sections 2(a)(7) and 2(a)(8) of the IUDTPA are sufficient

and, accordingly, denies 3DS's motion to dismiss Count VI.

VI. Tortious Interference with Prospective Economic Advantage (Count VII)

In Count VII, Desotech asserts a claim for tortious interference with prospective

economic advantage. To bring such a claim under Illinois law, a plaintiff must allege that "(1)

he had a reasonable expectancy of a valid business relationship; (2) defendant knew about the

expectancy; (3) defendant intentionally interfered with the expectancy and prevented it from

ripening into a valid business relationship; and (4) the intentional interference injured the

plaintiff." *Associated Underwriters of Am. Agency, Inc.* v. *McCarthy*, 826 N.E.2d 1160, 1169,

356 Ill. App. 3d 1010, 292 Ill. Dec. 724 (Ill. App. Ct. 1st Dist. 2005). Illinois law also requires

that the plaintiff's business expectancy be with a third party. *Ali* v. *Shaw*, 481 F.3d 942, 944 (7th

Cir. 2007) (citing *Schuler* v. *Abbott Labs*, 639 N.E.2d 144, 147, 265 Ill. App. 3d 991, 203 Ill.

Dec. 105 (Ill. App. Ct. 1st Dist. 1993)).

3DS contends that Desotech has not sufficiently alleged the first element of the cause of

action, a reasonable expectancy of a valid business relationship. 3DS cites *MJ & Partners*

Restaurant Ltd. Partnership v. *Zadikoff*, 126 F. Supp. 2d 1130 (N.D. Ill. 1999), for the

proposition that to establish the existence of a reasonable expectancy, Desotech must show that it

had more than the "mere hope" of continuing its relationship, *id.* at 1138, and asserts that "[f]or

three of the customers, Lockheed, Tangible Express and AP Proto, Desotech has merely alleged

24

facts showing that it hoped to sell resins to those customers but the customers chose . . . to buy resins from [3DS] instead." Defs.' Mem. at 33. 3DS's arguments, however, rely on an unduly narrow reading of the complaint. For example, Lockheed allegedly expressed its desire to continue using Desotech resins but was coerced into an agreement to purchase resins exclusively from 3DS. Viewing those allegations in the light most favorable to the plaintiff, the court finds that Desotech has alleged facts sufficient to show that it had much more than a mere hope of continuing its relationship with Lockheed.

3DS also argues that "for the two remaining customers, Express Pattern and Dynacept, Desotech does not even allege that they stopped buying [resins] from Desotech" and that Desotech has thus failed to establish the third required element, a showing that defendants interfered with the expectancy. Defs.' Mem. at 33–34. Here, again, however, 3DS has taken an overly literal view of the allegations. For example, Desotech has alleged that as a result of 3DS's conduct, "Express Pattern can no longer use [Desotech's] new resin." Am. Compl. ¶ 59. The logical bridge between (a) Express Pattern's no longer being able to use Desotech resin and (b) its ceasing to buy such resin is exceedingly short and one that the court is compelled to cross on this motion to dismiss. Furthermore, Desotech's allegations regarding other customers, such as Lockheed, who allegedly "agreed [with 3DS] . . . to stop purchasing resins from Desotech," *id.* ¶ 66, also satisfy the third element.

3DS further argues that Desotech fails to "show the required element of malice because Illinois recognizes a competitive privilege that protects competitors from claims of tortious interference with prospective business advantage." Defs.' Mem. at 31. In an action for interference with prospective business advantage, a defendant may raise the competitor's privilege as an affirmative defense. *Soderlund Bros., Inc.* v. *Carrier Corp.*, 663 N.E.2d 1, 8, 278 Ill. App. 3d 606, 215 Ill. Dec. 251 (1995). That privilege "allows one to divert business from one's competitors generally as well as from one's particular competitors provided one's intent is, at least in part, to further one's business and is not solely motivated by spite or ill will." *Id.* at 8 (citing *Candalaus Chi., Inc.* v. *Evans Mill Supply Co.*, 366 N.E.2d 319, 51 Ill. App. 3d 38, 9 Ill. Dec. 62 (1977); Restatement (Second) of Torts § 768(1)(d) & cmt. b, at 40 (1979)). But as the court notes in *Candalaus*, a decision on which 3DS relies, the privilege does not apply to a defendant whose conduct "create[s] or continue[s] an illegal restraint of competition." 366 N.E.2d at 326–27 (quoting Restatement (Second) of Torts § 768, at 39)). The Seventh Circuit has similarly explained that the competitor's privilege does not apply where "circumstances indicate unfair competition, that is, an unprivileged interference with prospective advantage." *A-Abart Elec. Supply, Inc.* v. *Emerson Elec. Co.*, 956 F.2d 1399, 1404–05 (7th Cir. 1992) (citing *Fishman* v. *Estate of Wirtz*, 807 F.2d 520, 546 (7th Cir.1987)).

As discussed above, Desotech has sufficiently alleged claims against 3DS for attempted monopolization and unreasonable restraint of trade under the Sherman Act. Because the same factual allegations form the basis for Desotech's tortious interference with prospective economic advantage claim, there is, at a minimum, an issue of fact as to whether 3DS's conduct was protected by the competitor's privilege. 3DS's motion to dismiss Count VII is therefore denied.

CONCLUSION AND ORDER

Defendants' motion to dismiss [#29] is granted in part and denied in part. The motion is granted with respect to Counts I and II and denied as to Counts III through VII.

26

Plaintiff's tying claims, Counts I and II of the amended complaint, are hereby dismissed without prejudice. Desotech is granted leave to replead its tying claims by March 2, 2009.

The discovery stay previously entered pending resolution of defendants' motion to dismiss is hereby lifted. Discovery may proceed on all of the outstanding antitrust and patent claims, including the patent damages issues.

A status hearing and scheduling conference is set for March 31, 2009 at 9:30 a.m. In the meantime, the parties are directed to confer about—and, if possible, submit—an agreed proposed case management schedule for the antitrust claims (a scheduling order for the patent claims having already been entered on December 2, 2008). They are also directed to consider whether a settlement conference at this time would facilitate resolution of the case.

Dated: January 26, 2009   Enter:_____ *Joan N. Lefkow* _____
                                                      JOAN HUMPHREY LEFKOW
                                                      United States District Judge

27

6

Interfacing with the Nanofuture

Evolutionary processes that have shaped the living world have been the result of integrated selection, mutation, genetic drifts, isolation, and gene flow through gradualism and punctuated equilibrium resulting in successful extant species that have survived the onslaughts of time and space to further play their role in shaping the future. At the same time, several species have become extinct as they were unsuccessful and did not survive the evolutionary process.

Gradualism is selection and variation that happens gradually, which over a short period of time is hard to notice. Small variations take place in some individuals developing favorable traits that help the organism to adapt to its environment and survive, whereas the variation in some individuals is with less helpful traits resulting in their nonsurvival. With such a change that is slow, constant, and consistent, the population gradually changes over a long period of time.

In punctuated equilibrium, change comes in spurts. There is a period of very little change, and then one or a few huge changes occur, often through mutations in the genes of a few individuals. Mutations are random changes in the DNA that are not inherited from the previous generation, but are passed on to generations that follow. Although mutations are often harmful, the mutations that result in punctuated equilibrium are very helpful to the individuals in their environments. As these mutations are so different and so helpful to the survival of those who have them, the proportion of individuals in the population who have the mutation/trait and those who do not changes significantly in a very short period of time. The species changes very rapidly over a few generations, and then settles down again to a period of little change.[1]

The business of nanotechnology has also developed along similar lines, as certain businesses have evolved and are evolving by punctuated equilibrium, and several others with a longer evolution that have evolved and are evolving by gradualism, resulting in a history of "extant" businesses that will determine the future trajectory of nanotechnology and "extinct" businesses that will serve as a source of strategic learning for the initiation of timely transformative and iterative adaptive changes, through conscious design and reflective activities for survival in an anticipated future.

Whatever the evolutionary mechanism may be, the principles of "competitive exclusion" would simultaneously operate in aggressive market ecosystems resulting in "survival of the fittest."

Nanotechnology is now in its most difficult period of life—middle age; having experienced an exciting and romantic youth, it is now poised for an expectant ambitious future. Transiting through these anxious moments, nano-businesses including investors will have to delicately handle the "middle-age syndrome" by avoiding over-dwelling on the past, dredging up lost opportunities, and second-guessing decisions. On the other hand, there are several examples of successes and test cases to become guiding models, including supportive resources and institutions to inject enthusiasm and serve as reinvigorating inspirational means to confront the midlife doldrums and competitive challenges. It is time to assess the emerging scenarios, examine the consequences, and consistently challenge and re-question the conclusions to define a focused sense of business purpose and operative direction.

Nanotechnology has developed in verticals leading to technology-based businesses. Although technologies in some of the verticals may have developed significantly, they may not lead to successful businesses due to multiple reasons of difficulties in scale-up, costs, volume, and value in the temporal marketplace, inadequate skilled labor, production and test equipment, regulatory nonconformance in specific applications, etc. However, due to the all-pervasive nature of the "nano," lateral linking of these verticals could lead to partnerships that are mutually symbiotic and open up new and innovative business opportunities.

A few recent setbacks in the IP-led nanotechnology businesses illustrate some of the issues confronting nano-businesses.

Evident Technologies, Inc., one of New York State's first nanotechnology companies, filed for Chapter 11 bankruptcy protection in July 2009, after being sued for patent infringement by Invitrogen Corp., citing mounting legal fees associated with its defense. It had amassed nearly $1 million in legal fees—significant compared to its $3.8 million in assets.

Through bankruptcy, Evident and Invitrogen reached a deal. Evident acknowledged it infringed Invitrogen's patent "in certain narrow limitations." Evident also agreed never to perform life sciences work, which it had no designs of doing anyway. In exchange, Invitrogen dropped its lawsuit, and said it would not pursue further claims against Evident.

On March 31, 2010, a federal judge approved Evident's plan to exit bankruptcy. The company restructured its business and today it is a leading nanotechnology company specializing in the commercialization of quantum dot semiconductor nanocrystals.

Evident was expected to start collecting a grant worth $75,000 a month from the federal Defense Advanced Research Projects Agency (DARPA), according to bankruptcy papers.

Ownership of Evident changed because of the bankruptcy, although its largest shareholder remained the same: SOLA Ltd., a hedge fund in New York City. SOLA owned 73% of Evident, up from 14.5% of shares at the time of the filing, according to court papers. A group of creditors, led by SOLA, provided $2.7 million in financing for Evident throughout the bankruptcy.[2]

Evident had a portfolio of developed technologies in the field of thermo-electrics based on nano-sized semiconductor crystals, also called quantum dots, designed to generate electricity from the process heat given off at power plants. The products based on these technologies include LED (light-emitting diode) products, such as Christmas tree lights and traffic signals. The technology also has other applications. It can be used to make refrigeration units more efficient and eliminate the need for an alternator in vehicles, improving fuel efficiency.

Evident has a strong patent portfolio covering its developed technologies as it owns or has an exclusive license to over 65 issued, pending, and provisional U.S. and foreign filings. This patent portfolio encompasses various aspects of semiconductor nanocrystal technologies from synthesis to application. The Evident IP portfolio extends to applications such as optoelectronics, inks, paints, solar cells, personal care products, and optical switching. Additional applications are being developed and added to Evident's expanding IP portfolio. The company offers a range of teaming and licensing arrangements that make their expertise, technology, and IP available to partners to explore the use of quantum dot semiconductor nanomaterials, adapt and leverage their established technologies, bridge to next-generation materials systems, and expand their research or product options in a cost-effective manner for speedy commercialization. Table 6.1 presents some of the key issued backbone patents of Evident Technologies, Inc.

TABLE 6.1

Evident Technologies, Inc. Patents

U.S. Patent No.	Title of Patent	Date Issued
8050303	Laser based on quantum dot activated media with forster resonant energy transfer excitation	November 1, 2011
7804587	System for the recognition of an optical signal and materials for use in such a system	September 28, 2010
7785657	Nanostructured layers, method of making nanostructured layers, and application thereof	August 31, 2010
7777870	Method and system for the recognition of an optical signal	August 17, 2010
7754329	Water-stable semiconductor nanocrystal complexes and methods of making same	July 13, 2010
7723744	Light-emitting device having semiconductor nanocrystal complexes	May 25, 2010
7524746	High-refractive index materials comprising semiconductor nanocrystal compositions, methods of making same	April 28, 2009
7399429	III-V semiconductor nanocrystal complexes and methods of making same	July 15, 2008

Source: http://www.patentgenius.com/assignee/EvidentTechnologiesInc.html

After the emergence from Chapter 11 in March 2010, Evident Technologies has aggressively bounced back in business, offering its technologies and services to various companies through their partnering strategies by leveraging on its patent portfolio. For example, on May 4, 2011, Evident Technologies Corporation and Samsung Electronics Co., Ltd entered into a comprehensive patent licensing and purchasing agreement for Evident's quantum dot LED technology. This agreement grants Samsung worldwide access to Evident's patent portfolio for all products red to quantum dot LEDs from manufacture of the quantum dot nanomaterials to final LED production. This combination would accelerate the development of nano-enabled products and its speedy entry into the global markets.

In February 2009, Holiday Creations/Diogen Lighting, Inc. and Evident Technologies, Inc. announced the signing of an exclusive licensing and purchase agreement to enable a new type of LED to be commercialized in the seasonal lighting strand markets in the United States and Canada.

Evident Technologies, Inc. provides an example of how business adaptability supported by proprietary innovative products, solutions, and technologies backed by strong intellectual property position, coupled with strategic IP transactions, can navigate a company to traverse a torturous path and survive adverse conditions, and then recover via planned regenerative or mutative processes.

Luna Innovations, Inc., a Virginia-based nanomaterial company focusing on sensing, instrumentation, and nanotechnology, also filed for Chapter 11 bankruptcy protection in July 2009 after facing a jury verdict of $36 million in damages in a trade secret case—almost double its $20 million in assets. The case was filed by Hansen Medical, Inc. On January 12, 2010, it emerged from Chapter 11 reorganization. The Hon. William F. Stone, Jr. of the U.S. Bankruptcy Court for the Western District of Virginia, Roanoke Division, confirmed the company's Joint Plan of Reorganization on January 12, 2010. Luna's reorganization plan provided that Luna's creditors would receive a 100% recovery on their valid claims and that Luna's current shareholders would retain their shares. Hansen Medical, Inc. and Luna Innovations, Inc. reached a settlement in December 2009, resolving the outstanding litigation between them. The settlement resulted in a development and supply agreement between Luna and Hansen, and a license of Luna's fiber optic shape sensing technology to Hansen in the fields of medical robotics and certain medical nonrobotics.[3]

In January 2009, QuantumWise bought all assets of Atomistix after it went bankrupt. Atomistix was a provider of software solutions for development of nanotechnology. Since its incorporation in Copenhagen (2003), the company has been working in close collaboration with the Nano-Science Center at the Niels Bohr Institute of Copenhagen University.

The textile manufacturer Burlington Industries, which filed for bankruptcy in 2001, was bought in 2003 by International Textile Group (ITG) for more than $600 million, largely because it owned a controlling interest in Nano-Tex, an Emoryville, California, company with a patent on stain-resistant material, which was made possible by recent nano advances. Nano-Tex-based products are able to command a 30% premium at retail because of their unique properties. This made Burlington's interest in Nano-Tex worth a fortune and may be the start of a trend for embattled manufacturing companies.[4]

In April 2010, Sunfilm AG filed for insolvency at the district court in Dresden, Germany, as its shareholders had stopped their financial support. The company cited unfavorable market conditions and feed-in tariff discussions in Germany for market entry. Its competitor, Oerlikon Solar's IP Portfolio included European Patent EP 0 871 979 B1, which claimed micromorph® tandem cell technology. In 2003, Oerlikon had obtained an exclusive license from IMT, University of Neuchatel (Switzerland). Oerlikon had filed a patent infringement suit against Sunfilm AG. However, the patent was invalidated in May 2009.

Despite this victory by Sunfilm AG, it could not sustain its business due to market pressures and decided to explore strategic realignment of the company with a new investor with the belief that it had high-performance products and state-of-the-art production lines in combination with an above-average market, growth potential of their technology, and competence of their employees.

Optiva, a nanotech company founded in 1997, attracted $41.5 million venture capital funding including investments from JP Morgan Partners for its laminated flat-screen TV sets. It had to shut down in April 2005 after it failed to continue to raise further funds. The problem was that it took too long to release its product, which was obsolete by the time it came to market.

In December 2011, Berlin-based solar panel manufacturer Solon also became a victim of a wrenching consolidation process and brutal price competition. It filed for bankruptcy, as it was unsuccessful in restructuring its loans with banks. It sought to restructure those loans for three of its subsidiaries through insolvency proceedings. The sustainability of companies such as Phoenix Solar, Q-Cells, and Conergy has also been questioned.

In the United States, the once-successful solar panel maker Evergreen Solar filed for bankruptcy and sold off its assets in 2011. Less-established companies Solyndra, SpectraWatt, and Stirling Energy Systems also went out of business in 2011.

The primary reason for these failed businesses in the solar field is low-cost competition led by Chinese manufacturers. It is estimated that panel prices have fallen more than 50% in the last 3.5 years. Industry analysts say that more solar companies, including those in China, will either be bought or shut down, as solar panels have become commodities. Government subsidies for the solar sector are also at risk, making the demand picture unclear in the years ahead.

6.1 Uncertainties Looming over IP Portfolios

In a field such as nanotechnology that is already overcrowded with patents and in several cases with overlapping claims, commercialization of nano-enabled products and services faces potentially costly and complex patent litigations. Successfully navigating through the dense nanotechnology patent landscape will require in-depth surveying of undulations in the terrain, booby-traps, and patent mines that could adversely impinge on the commercialization of "nanovations." Clearing patent obstructions or working around them will require strategic planning with a clear understanding of the techno-legal issues in this domain. In the future, there will be intense oppositions in pregrant proceedings (where the law allows for pregrant opposition as in India) or postgrant oppositions (where the law allows for postgrant opposition), or revocation proceedings including re-examination procedures as in the United States to clear or weaken the threat of blocking patents.

A nanotechnology company's commercial success will depend, in part, on its ability to obtain and maintain effective patent and other protection for the technologies underlying the materials, processes, and applications relating to the company's products and to successfully defend patent rights in those technologies against third-party challenges.

There can be no assurance that any of the patent applications filed will be granted or, if granted, will be enforceable. The claims as filed may get narrowed due to amendments carried out during prosecution of the patent application and/or during opposition/re-examination/revocation/validity proceedings.

It ought to be appreciated that if granted patents are invalidated, they will fail to provide any competitive advantage. Further, in a fierce competitive environment, the competitors may develop competing technologies, thereby impacting the value of any existing patent portfolio of a company.

In a technology-led field such as nanotechnology, a company may receive cease and desist notices from competitors alleging that the company is infringing on some of the competitor's patents or other IPs. It should be understood that a company may incur substantial litigation costs, whether or not the litigation is determined in its favor. The company may also have to negotiate licenses of the relevant IPs of other parties for the business continuity. The company would then have to strategize its litigation strategy to avoid significant legal expenses, an adverse effect on sales and diversion of efforts of its scientific and management personnel.

Protection of a company's confidential information and know-how is of immense significance especially related to the competitive technologies and knowledge it possesses that may or may not be patentable. For strategic reasons, a company may decide not to patent its innovations and maintain them as trade secrets. It is imperative that the company has a well-defined policy that is systematically implemented so that it can assert its rights on such confidential information/know-how.

In summary:

Patents claims may not be sufficiently broad in their scope to provide adequate space to enforce one's patent rights against third parties.

There can be no assurance as to the ownership, validity, or scope of any patents that have been, or may in the future be, issued to a company or that claims with respect thereto would not be asserted by other parties.

Substantial costs may be incurred if a company challenges the proprietary rights of others or is required to defend its proprietary rights.

Commercial success of a company would depend upon noninfringement of patents granted to third parties that may have filed applications or that have obtained or may obtain patents relating to products that might inhibit the company's ability to develop and exploit its own products. If this is the case, the company may have to obtain alternative technology or reach commercial terms on the exploitation of other parties' intellectual property rights.

There can be no assurance that a company will be able to obtain alternative technology or, if any licenses are required, that the company will be able to obtain any such license on commercially favorable terms, if at all. This may have an adverse effect on the company.

A company may in-license certain technology and the right to use certain patents that are owned by third parties. Patent applications filed in a number of countries with respect to these in-licensed patents, and in certain circumstances, the patents may be granted but there can be no assurance that any such applications will be granted or, if granted, be enforceable, and they may be amended to reduce the scope of protection of patent claims. In addition, there can be no guarantee that patents the company in-licenses are valid and not open to challenge. In addition, failure to develop products based on technology licensed by the company could result in the unexploited license being terminated.

There is an emergence of enterprises termed "patent trolls" that are not actually producing a tangible product or service but are merely trading in a secondary market of patents for which others have filed but have not enforced for a number of reasons, such as a lack of funds or when the originating company has gone bankrupt and the patents have been sold as a part of the realization of the assets of the company for its debtors. Such enterprises have also been called "nonpracticing entities" (NPEs). For such enterprises, the potential royalties, infringement damages, and recoverable switching costs drive their business model. In one of the activities of a patent troll, it acquires intellectual property to create a strategic portfolio, and then waits for the most appropriate opportunity to sue a potential infringer, to either negotiate terms with the company (alleged infringer) or file infringement suits to derive royalty/damages. In several instances, although patent trolls become hurdles to commercialization of innovations, these trolls also act as a single window to identify inventions for licensing. The nanotechnology-patenting scene although fragmented is increasingly being populated with patents with overlapping claims and complex ownerships. Further, there are instances of several entrepreneurs, universities, and centers of excellence, especially in developing and least developed countries, being unable to progress their patent applications due to lack of funds, facilities, and resources leading to "orphaning of inventions," thereby leading to the creation of fertile grounds for the germination and growth of patent trolls that are engaged in hoarding of inventions with their associated intellectual property rights—a trend that is disturbing and undesirable as they could become barriers to innovation and local technology development.

6.2 Case Study: Oxonica Ltd. (formerly Oxonica plc)

Oxonica Ltd. is an example of a business that remodeled its businesses from time to time by growing technology-led companies with rich patent portfolios, striking strategic alliances with other technology-led companies, and then selling off its subsidiaries to meet its financial objectives. The company also presents an example of the fate of a well-established business that can get into financial problems due to patent infringement disputes. The company is further restructuring its business and derives its revenues from the royalties of its licensed technologies and patents. This case study also illustrates the importance of well-structured contracts and agreements as errors in the licensing/IP transaction agreements can be fatal to a company's future as has been the case between Oxonica and Neuftec Ltd. as is summarized in this chapter and Chapter 5.

Oxonica was a spin-off from Oxford University, England in 1999 after seven years background research originally to make and sell nanoparticles for display applications. It created a business for advanced nanomaterials with its headquarters in Haddenham, U.K. focused on energy, environment, and healthcare especially via sunscreens, fuel additives, and biodiagnostics through tailoring nanoparticles for customer applications, building revenues based on IP generation. It obtained funding of £2.3M from Angels and DTI awards and £8.2M from institutional funding.

In 2005, while floating on the Alternative Investment Market (AIM), London Stock Exchange, the Oxonica Group had a dense proprietary technology landscape with a patent portfolio of 29 families and 10 granted patents with further applications in 29 families filed and pending. The Group had filed trademark applications in various jurisdictions for Oxonica, Envirox, Optisol, Serrplex, Serrcode, Cerulean, and the Cerulean logo. The IPR portfolio and strategy were managed by the internal Intellectual Property Strategy Committee (IPSC) to align with the business strategy of Oxonica Group. The IPSC ensured that the Group conducted periodic searches for patents at

TABLE 6.2

Patents Covering Group Activities

Field	Number of Patents Granted	Number of Patent Families
Fuel additives	1	4
Optisol	3	7
Biodiagnostics	2	9
Polymers & coatings	0	3
New product development	4	6

grant or application stage to identify any potential third-party infringements and, wherever appropriate, Oxonica obtained infringement opinions from its retained patent attorneys. The patents covering all the areas of the Group's activities are summarized in Table 6.2.

The company subsequently added to this portfolio several patents/patent applications and further expanded to technologies related to optical security tags.

To supplement in-house patents, the Group had a strategy of in-licensing certain technologies and building IP positions around the in-licensed technology. These included seven patent families that included the core Optisol Technology (titania nanoparticles used in sunscreen to give UV protection and safeguard against free radical damage), which was licensed exclusively to the Group by Isis Innovation Ltd, Oxford University's IP Exploitation and Technology Transfer Company. The Oxonica Group also filed several patent applications based on their own R&D.

Oxonica Energy had in-licensed one core patent family underpinning the fuel catalyst technology from Neuftec Limited for the Envirox fuel-borne catalyst (nanoparticles of cerium oxide used in diesel fuel at ~5 ppm to eliminate soot and improve fuel efficiency), had access to four patent families covering biodiagnostic applications in-licensed from University of Strathclyde, and had one application developed as part of the joint project with Johnson Matthey plc filed in Oxonica Energy's name. The R&D efforts in Oxonica Energy were directed to a method to improve the dispersion and settling properties of the catalyst in diesel fuel. The Neuftec license was exclusive and runs until this patent expires (on June 29, 2021) or is revoked or the license is terminated according to its terms set out in Part II of the agreement. The license was based on a combination of royalty and milestone payments and profit share.

A chronology of events tracks the Oxonica Group's business trajectory over the years and there are several lessons to learn for the future.

- 1999: Spin-off from Oxford University.
- 2002: Became revenue generating.
- July 2005: Floated on Alternative Investment Market (AIM), London Stock Exchange with market cap of £35M.

- December 2005: Took over Nanoplex (US). Oxonica signed an agreement to acquire U.S.-based Nanoplex Technologies, Inc. for a total of up to 7,538,440 fully paid ordinary shares in Oxonica, which is equivalent to approximately 17% of the equity of the enlarged group. Nanoplex changed its name to Oxonica, Inc. and became part of Oxonica's healthcare business. The acquisition enhanced the development of Oxonica's biomarker detection and security technology including products based on customer needs in high-growth diagnostics markets, such as healthcare, homeland, and brand security. Oxonica Plc operated in three divisions, namely Oxonica Energy, Oxonica Healthcare, and Oxonica Materials.

- 2007: Deal with a Turkish oil company broke down and its valuation was reduced.

- 2007: Transferred biodiagnostics to U.S. operation.

- October 13, 2008: Judgment by Deputy Judge Peter Prescott QC in the High Court concerning the interpretation of the license agreement between Oxonica Energy and Neuftec and ordering the payment of royalties to Neuftec on sales of Envirox™ up to the date of termination of the license agreement. Oxonica Energy had been granted permission to appeal this decision. An initial amount of £394,000 covering the royalties in dispute was placed by Oxonica Energy in a joint bank account of both parties' solicitors in October 2008.

- June 23, 2009: Decision from the Court of Appeal U.K. on the long-standing patent license dispute between Oxonica Energy Limited ("Energy") and Neuftec Limited ("Neuftec"). The court upheld the earlier judgment of Deputy Judge Peter Prescott QC. An additional amount of £550,000 was to be paid further to the £394,000 royalties following the High Court hearing. Oxonica had cash balances of £1.6 million as of July 7, 2009.

- August 4, 2009: Oxonica plc was de-listed from AIM.

- September 2009: Oxonica Energy Limited was sold to Energenics in full and final settlement of legal actions between Neuftec and Oxonica.

- September 2009: Oxonica plc signed a strategic agreement relating to its Nanoplex™ technology with BD (Becton, Dickinson and Company), a leading global medical technology company headquartered in New Jersey. This agreement follows from the license agreement signed with BD in August 2006 and subsequent technology development agreements. Under the terms of the assignment and license agreement, BD would pay Oxonica a total of up to $7 million. Of that, $3.5 million was payable on signing the agreement in exchange for the assignment to BD of the patents covering Oxonica's Nanoplex technology, related know-how, and the

Nanoplex trademark. A further $3.5 million was payable on Oxonica completing certain technical transfer milestones. In addition, BD would make payments to Oxonica on sales of BD products covered by the Nanoplex patents; such payments would replace the original license agreement. Because of this transaction, BD would assume the responsibility and costs associated with prosecuting the patent portfolio and would take responsibility for the manufacture of the Nanoplex tags for its own requirements. This would allow Oxonica to achieve a significant reduction in operating expenses. Oxonica was granted a royalty-free, exclusive license to continue to use the technology in the fields of industrial and homeland security and to continue to evaluate opportunities in *in vivo* diagnostics, agriculture (excluding veterinary), and fine chemicals.

- September 2009: Oxonica concluded an agreement with Croda plc, under which it has assigned all patent rights, trademarks, and know-how associated with Optisol™ and Solacor™ technologies for applications in personal care and polymers. Croda will exclusively market and distribute the products to all global markets and pay royalty fees on sales to Oxonica.

- July 2010: Oxonica announced the sale of its subsidiary company, Oxonica Materials, Inc. (OMI), to Cabot Corporation in an all-cash transaction of approximately $4 million. OMI is a leading developer of surface-enhanced Raman scattering (SERS) materials and detection methods. These SERS materials provide a unique signal that can be detected using specialized readers and therefore provide highly counterfeit-resistant security solutions for a broad range of applications including brand security, fuel markers, tax stamps, and identification. Cabot Corporation's security business is a leading supplier of covert taggants for many high-security applications.

- February 11, 2011: Oxonica Plc was re-registered as Oxonica Ltd., a private limited company.

The company will now have to restrategize and restructure its operations to ensure its sustainable future.

6.3 United They Stand, Divided They Fall

The saying "united we stand, divided we fall" is a parable that is so true to the future of nanotechnology. To enhance an organization's innovation capability and commercial viability in nanotechnology businesses at this point, resources will have to be unified through the formation of strategic

alliances, managing dynamic exchange of information and active relationship building with external research organizations for productive collaborations, marketing networks, technology/applications and business
integrators, and communicators, all managed through unambiguous contractual agreements. A company may also acquire other companies that
include the technology, skill bases with working teams, and the licensing of
associated technologies, software, tools, and designs. A company may also
develop complementary new stream activities via mergers, acquisitions,
joint ventures, and precompetitive collaborative developments that enhance
innovation and simultaneously broaden a company's business portfolio. The
future of nanotechnology, especially due to the complex IP scenario coupled
with competitive market forces and emerging regulatory issues involving
risk profiling in health, safety, and environment, is expected to be a continuous game of "tug-of-war." Unproductive war games consume resources
and, therefore, with each technological advance, the fighting must stop and
agreements must be reached between the stakeholders for mutual benefits
and successful commercialization of nanotechnology.

There are various modes by which organizations share their intellectual
property rights. In unilateral modes of licensing that may include exclusive
or nonexclusive licenses, the parties (licensor and licensee) may negotiate to
share some risk due to an uncertain patent landscape.[5]

In various areas of nanotechnology, there are situations when several patent owners owning overlapping patent rights separately demand royalty
from anyone using their technologies to bring a product to the market. The
net effect of such multiple demands of royalties is royalty stacking. One may
negotiate an antistacking provision in the licensing agreements to prevent
royalty stacking; a provision that would require the licensing patent holder
to share some of the financial burden by reducing the royalty rate by a percentage payable to the patent owner if other third-party licenses are required
for the same product.

It is not uncommon to include an indemnification clause in a licensing
agreement. This is another mode by which the licensor shares the risk of an
uncertain patent landscape by agreeing to defend the licensee from patent
infringement claims by third parties.

Cross licensing is yet another mode of IP transaction by which competing companies clear blocking patent positions among themselves through
mutual sharing of their respective patents. In this mode, the parties grant
each the right to practice the other's patents. In several cases, the cross licensing involves an entire portfolio of patents from the participating parties.
Such transactions may or may not involve monitory payments. However, if
the patents are held by parties whose business interests do not converge,
cross-licensing does not serve as a working model. In the case of nanotechnology's multidisciplinary nature, the relevant patent holders may not be
competitors in the same sector and, therefore, may have little interest in
exchanging patents.

For example, in a cross-licensing patent deal on January 10, 2011 between Intel and Nvidia, Intel will gain access to Nvidia's patents while paying Nvidia $1.5 billion in licensing fees as part of a new six-year agreement. For the future use of Nvidia's technology, Intel will pay Nvidia an aggregate of $1.5 billion in licensing fees payable in five annual installments, which began January 18, 2011. Nvidia and Intel have agreed to drop all outstanding legal disputes between them. Intel and Nvidia had sued each other in early 2009 in a dispute that originally centered on a chipset license agreement. Intel had contended the cross license does not extend to Intel's future-generation processors, and Nvidia countersued blocking access to its patent portfolio. The crux of the agreement is that Intel gains access to all of Nvidia's GPU (graphics processing unit) patents but Nvidia gains access only to certain Intel patents. To compensate for the lopsided patent access (which favors Intel), Intel pays Nvidia $1.5 billion. The agreement still bars Nvidia from using any of Intel's x86 technology and, as a result, Nvidia cannot build x86-compatible chipsets. In addition, many PC makers (including Apple) still use discrete (standalone) Nvidia GPUs that attach to Intel chipsets. Intel is paying for the GPU patents that Nvidia incorporates into its ARM processors. GPUs excel at parallel processing, whereas CPUs (central processing units) such as Intel's x86 chips do sequential processing. Both types of processors have their merits, although GPUs have the potential to be much faster than CPUs at doing visual processing and scientific number crunching, for example. Nvidia will continue its focus on ARM processors, which compete with Intel's x86 chips in small devices like Netbooks and tablets.[6]

As an alternative to cross licensing, "Patent Pools" (PPs) may be formed by diverse patent owners who feel the need to use each other's essential patents in a fragmented patent landscape to develop a product., As a result, they pool their essential patents in a common basket (functioning as a clearinghouse) to function as a one–stop shop, to enable each pool member to access the pooled patents with distributed risks in accordance with the common agreement reached between the parties while forming the PP. Such PPs, if administered properly, could clear the hurdles created by "blocking patents" and "license staking" and significantly reduce the cost of licensing transactions and litigations.

To create and administer such PPs, the parties must

- identify the essential patents owned by each party with respect to the technology;
- agree on an assigned value to each such essential patent and the framework for distributing the royalty dividends;
- grant nonexclusive licenses to the pool, thereby maintaining a proviso that the pool members are also free to license their patents outside the pool;
- agree on the overall royalty rate, along with other terms under which the patents in the pool will be licensed to interested parties in a nondiscriminatory manner;

- all grant back provisions are limited to essential patents and require nonexclusive licenses with fair and reasonable terms so as not to inhibit or discourage further innovations;
- ensure that the pool does not violate antitrust rules/competition laws in the jurisdictions they operate and is viewed as a patent cartel, a means of shielding patents that may be invalidated, a collusion, a common price fixing body, etc.[7]

Nanobased activities have grown rapidly in the past decade. Several institutions have been set up (governmental, nongovernmental) to facilitate nanotech companies in a variety of ways. Various facilitating companies and institutions are listed in Table 6.3.

6.3.1 Case Study: Dendrimers and Commercialization by Starpharma Holdings Limited

The development of dendrimers and the path to commercialization is a representative illustration of the complexities to be tackled to set the foundations of a possible successful nanotechnology business of the immediate future.

A dendrimer is generally described as a macromolecule, which is characterized by its highly branched 3D structure that provides a high degree of surface functionality and versatility with perceived applications in medical, electronics, chemicals, and materials industries. These macromolecules were first synthesized in 1978, and a refined methodology was put in place from about 1979. The term "dendrimer" was first used in U.S. Patent No. 4,507,466, which was filed on January 7, 1983 in the names of Donald A. Tomalia and James R. Dewald (listed assignee: Dow Chemical). Its term expired on January 7, 2003. In a broad U.S. Patent 5,527,524, June 18, 1996 titled "Dense Star Polymer Conjugates" issued in 1996, Tomalia and colleagues claimed dendrimer polymers as complexing and carrier materials for biological agents, including genes, DNA for transfection, IgG and such agents as the herbicides 2,4-dichlorophenoacetic acid, and abscissic acid. Another early U.S. Patent 5,714,166, of Tomalia and colleagues with Dow Chemical as assignee, issued on February 3, 1998 title "Bioactive and/or Targeted Dendrimer Conjugates" is of significance. Since then, several patents have been filed and issued in dendrimers spanning diverse fields of application by various applicants.

From 1985 to 1995, Dow Chemical and Stamford, Connecticut–based Xerox were dominant dendrimer patent holders, the latter patenting the use of dendrimers in toner and ink dispersion. Bayer and DSM of Heerlen, the Netherlands, also were granted patents for the use of dendrimers in plastics manufacturing and other nano-based products.[8]

In 1992, Tomalia left Dow Chemical to found Midland, Michigan–based Dendritech, which was 90% owned by the Michigan Molecular Institute, a nonprofit institution and 10% owned by Dow Chemical. In August 1996,

TABLE 6.3

Starpharma Patent Portfolio

Title	Priority Date; Publication Number	Patents Granted	Patents Pending
VivaGel® Patent Portfolio			
Antiviral Dendrimers	June 15, 1994 WO95/34595	Australia, Brazil, Canada, China, Europe, Hong Kong, Mexico, New Zealand, Singapore, South Korea, U.S., Japan	
Anionic or Cationic Dendrimer Antimicrobial or Antiparasitic Compositions	September 14, 1998 WO00/15240	Australia, Canada, Europe, Mexico, New Zealand, Singapore, South Korea, U.S.	China, Japan
Agents for the Prevention & Treatment of Sexually Transmitted Diseases-I	March 30, 2001 WO02/079299	Australia, Canada China, Europe, Hong Kong, Japan, Mexico, New Zealand, Singapore, South Korea, U.S.	Brazil, U.S.
Microbicidal Dendrimer Composition Delivery System	October 18, 2005 WO2007/045009	New Zealand, Russian, Federation	Argentina, Australia, Canada, China, Europe, Hong Kong, India, Japan, Malaysia, Mexico, South Korea, Taiwan, U.S.
Contraceptive Composition	March 22, 2006 WO2007/106944		Australia, Canada, China, Europe, Japan, U.S.
Method of Treatment or Prophylaxis of Bacterial Vaginosis	May 16, 2011 Not yet published		International application
Platform Patent Portfolio			
Macromolecules Compounds Having Controlled Stoichiometry	October 25, 2005 WO2007/048190		Australia, Canada, Europe, U.S.
Modified Macromolecules	August 10, 2006 WO2007/082331		Australia, Canada, China, Europe, India, Japan, U.S.

(Continued)

TABLE 6.3 (*Continued*)

Starpharma Patent Portfolio

Title	Priority Date; Publication Number	Patents Granted	Patents Pending
Dendritic Polymers with Enhanced Amplification and Interior Functionality	April 20, 2005 WO2006/065266	Canada, India, Japan, New Zealand Singapore, South Korea, U.S.	Argentina, Brazil, China, Europe, Hong Kong, Israel, Mexico, Taiwan
Dendritic Polymers with Enhanced Amplification and Interior Functionality	December 21, 2005 WO2006/115547	Australia, India, Singapore, South Korea, U.S.	Argentina, Brazil, Canada, China, Europe, Hong Kong, Israel, Mexico, New Zealand, Taiwan
Imaging Project Patent Portfolio			
Polylysine Dendrimer Contrast Agent	August 11, 2006 WO2008/017122		China, Europe, U.S.
siRNA Project Patent Portfolio			
Delivery of Biologically Active Materials Using Core-Shell Tectodendritic Polymers	March 3, 2006 WO2008/054466		Europe, U.S.
Drug Delivery Project Patent Portfolio			
Targeted Polylysine Dendrimer Therapeutic Agent	August 11, 2006 WO2008/017125		China, Europe, India, U.S.
Agricultural Chemicals Patent Portfolio			
PEHAM Dendrimers for use in Agriculture	October 26, 2009 WO2011/053605		International, U.S.

Source: Starpharma Annual Report, 2011, http://www.starpharma.com/assets/downloads/ annual_reports/Annual_Report_2011_Web.pdf

DSM reached an agreement with Dendritech to license a number of patents relating to dendrimers and their applications owned by Dendritech including "pamam" (polyamidoamine) dendrimers based on ethylene diamine and methyl acrylate raw materials as well as patents licensed to Dentritech by Dow Chemical. By 1996, Dendritech had earned licensing fees of $20 million from firms experimenting in drug delivery, gene therapy, and personal care products. Firms that had concluded deals with Dendritech included Procter & Gamble, Unilever, Abbott Laboratories, and L'Oreal.

Dentritech sold its dendrimer patents back to Dow Chemical in 2000. Then Tomalia left Dendritech and founded a new company, Dendritic NanoTechnologies, Inc. (DNT). In January 2005, Dow turned over its entire IP portfolio on dendrimers (196 patents worldwide) to DNT in exchange for owning a significant stake in the company.

Starpharma Pooled Development Ltd. (Australia) was an ASX listed company established in 1996 to develop polyvalent nanoscale molecules (dendrimers), which had been under development by the Biomolecular Research Institute (BRI) in Melbourne, Australia, since 1992. The BRI first investigated dendrimers as protein mimics for pharmaceutical applications, before licensing the technology to Starpharma Pooled Development Ltd.'s subsidiary company, Starpharma Ltd. ("Starpharma"). Funding for Starpharma's development projects has been sought and obtained from government (Australian and U.S.) grants. In this way, shareholder equity is maintained, and even increased, compared with the alternative strategy of raising additional capital through the issue of new shares. A key part of Starpharma's strategy in reducing development cost is to form relationships and alliances with leading companies and institutions around the world, including universities, contract research organizations, and contract manufacturers. Starpharma's extensive network of international collaborators significantly expanded its capabilities in the full spectrum of drug development activities from discovery research through to GMP manufacture.[9]

In October 2006, Starpharma (Melbourne, Australia), which already held a 42% interest in DNT, acquired Dendritic Nanotechnologies for $6.97 million in shares and headquartered the company in Melbourne with production and laboratories in Central Michigan University's Center for Applied Research and Technology in Mount Pleasant. DNT became a wholly owned operating subsidiary of Starpharma Holdings, with Dow owning approximately 8.6% of Starpharma. DNT had more than 30 patents in dendrimer science, and sold and licensed more than 200 variations of dendrimers to pharmaceutical, biotechnology, and diagnostics companies.

This acquisition made Starpharma the dominant dendrimer patent holder with complete access to DNT's patents and know-how of its Priostar dendrimers, with existing royalty streams in place from leading life-science companies. This gave Starpharma the opportunity to commercialize dendrimer technology not only in the pharmaceutical sector but also into other nearer-term life science and industrial applications.[10] The deal also positioned DNT and Starpharma as the leading out-licensing sources of license rights for dendrimers and as developers of the technology in their own rights. This move consolidated a great amount of the important intellectual property in the dendrimer field into one company that would have a very positive impact for developing the applications and further demonstrating the value of this technology and thereby positioning Starpharma with Dow's expertise of the provision of diversified product pipeline with near-term cash-flow opportunities, and a more balanced risk profile to develop, market, and successfully commercialize these technologies. This deal also gave Starpharma an increased U.S. presence.

It may be recalled that in January 2004, Starpharma became the first company in the world to initiate human clinical testing of a dendrimer-based pharmaceutical (VivaGel™ for prevention of HIV) under a U.S. Food and Drug Administration Investigational New Drug application.

In March 2005, Starpharma and Industrial Research Ltd., New Zealand (a commercial company with the New Zealand government as the shareholder) had established a joint venture to accelerate the development of carbohydrate-functionalized dendrimers (glycodendrimers) for use in therapeutics, antigen presentation, and as biologically active compounds. This coincided with a $NZ945,000 grant from the Australia New Zealand Biotechnology Partnership Fund to support the joint venture activities. Industrial Research would invest a further $NZ1.02 million in the field to position New Zealand industry as a leader in the development and commercialization of glycodendrimers for pharmaceutical applications. The joint venture would select a series of lead candidates for further development. Starpharma would take a lead role in the development and commercialization of products with Industrial Research, and Industrial Research's cGMP* manufacturing business unit, GlycoSyn, would provide manufacturing and specialized expertise in carbohydrate design, synthesis, and analysis.[11] In February 2007, Starpharma, through its U.S. subsidiary company Dendritic Nanotechnologies, Inc. (DNT), entered into a worldwide exclusive license and supply agreement with EMD Biosciences, part of Merck KGaA's Performance and Life Science Chemicals division. Under the terms of this agreement, DNT would supply EMD Biosciences with PrioFect transfection reagents based on Priostar proprietary dendrimers for the DNA and siRNA transfection research markets. Terms of the agreement, which included royalties and milestone payments, were not disclosed.

DNT's PrioFect transfection reagents are part of the $200 million market for nucleic acid, DNA, and small interfering RNA (siRNA) research. PrioFect transfection reagents are the only transfection reagents with nanometer-size control, enabling EMD to offer researchers siRNA transfection reagents with sizes optimized for individual cell lines. Under this commercial arrangement, DNT retains full rights to all *in vivo* aspects of transfecting nucleic acids with Priostar technology, a market segment that experienced significant deal-making activity in 2006.

The license and supply agreement with EMD Biosciences, the first since Starpharma acquired DNT, is significant because it will lead to the first commercial application of Priostar dendrimers.

The agreement introduced Starpharma as a player in siRNA research, an area that is undergoing rapid growth and seems poised to become a major source of new medicines for many human diseases.

Through the collaboration between EMD Biosciences and DNT, they would provide leading edge technology for a rapidly growing transfection reagent market segment and utilize the technology as a foundation for future product platforms and also develop unique and highly efficient transfection reagents that would be marketed through the Novagen brand of products.

In July 2011, Starpharma announced that its studies have demonstrated a number of improvements in these preliminary studies including the ability to increase the effectiveness of agrochemicals, such as glyphosate, the most commonly used herbicide globally (also known by the trade name Roundup®) with annual sales in excess of $5 billion. In the past year, Starpharma's internal agrochemical program has explored a number of key off-patent agrochemical agents in combination with the company's proprietary dendrimer technology. In addition to glyphosate, these include the major insecticide imidacloprid (annual sales of $1 billion) and the herbicide trifluralin (annual sales of $300 million). Based on initial studies, Starpharma's Priostar® dendrimers are well placed to capture several opportunities in the $40 billion global agrochemical sector as the market continues to seek new technologies to improve efficiency and enhance performance.

Starpharma's internal agrochemicals program focuses on reformulating known generic agents with its proprietary dendrimers to improve their performance. This offers the potential for reduced frequency and amount of application, with the potential to reduce the chemicals' environmental impact.

Key patents have already been allowed or granted by the United States and other patent offices for broad protection of Priostar® dendrimer technology, relevant to both agrochemical and industrial applications. Additionally, Starpharma has filed for protection for specific agrochemical applications, which if granted would provide patent coverage to 2029.

Starpharma's agrochemicals program has been assisted by funding under the Victorian Government's STIUP program announced in March 2011. Starpharma also has partnered programs with a growing number of leading agrochemical companies.

In August 2011, GlaxoSmithKline (GSK) and Starpharma announced that GSK was awarded a grant to advance a dermal treatment based on Starpharma's dendrimer drug delivery technology. The funds will be used to support Starpharma's synthesis of dendrimer-based drug candidates, which will then be tested by Stiefel, a GSK company with a view to further development toward a dermal product. The funding has been provided under a grant program run by the Victorian Government.[12]

The Starpharma patent portfolio currently has around 30 active patent families with over 110 granted patents and more than 70 patent applications pending. Two new provisional patent applications were filed during the year. Key patents within the Starpharma portfolio as of August 9, 2011 are provided in Table 6.4.

The corporate alliances/partnerships of Starpharma over the years have been established to explore dendrimers' potential applications such as new drugs with unique activities, targeted drug delivery, or modifying agents for

TABLE 6.4

Networking and Facilitating Development and Commercialization in the Nanotechnology Sector

Helmut Kaiser	Provides exclusive international strategies, schemes, technology/market studies, and special studies for concerns, small and midsized enterprises, and government agencies.
NanoKTN	Nanotechnology Knowledge Transfer Network (NanoKTN) is a network to promote nanotechnology business and simplify the nanotechnology innovation landscape by providing a clear and focused vehicle for the rapid transfer of high-quality information on technologies, markets, funding, and partnering opportunities.
Go BIG Network	Online network of small businesses, start-up companies, and investors.
Nano Science and Technology Consortium	Nongovernmental, industry-managed and -promoted organization with a role of facilitator for nano developmental activities.
CC-NanoChem	German network for innovative nanotechnology-based materials offering associated enterprises, research institutes, founders, and investors a common platform for cooperation, exchange, further education, and advanced training and contact with the public processes.
Bio Life Technical	Provides due diligence services in nanotechnology and personalized healthcare industry to global investors.
Applied Materials Ventures	Focuses on electronic and photonic components, systems, subsystems, and software for datacom and telecom applications as well as nanotechnology.
SEMATECH	Provide services to its member companies to address issues related to the timely availability of the materials, tools, and technology to stay on the International Technology Roadmap for Semiconductors (ITRS).
University Innovation Centre for Nanotechnology	The National Institute for Nanotechnology (NINT) is an integrated, multidisciplinary institution involving researchers in physics, chemistry, engineering, biology, informatics, pharmacy, and medicine. Established in 2001, it is operated as a partnership between the National Research Council and the University of Alberta, and is jointly funded by the Government of Canada, the Government of Alberta, and the university.
In-Q-Tel	Independent, private, not-for-profit company to help the CIA and the greater U.S. Intelligence Community (IC) to identify, acquire, and deploy cutting-edge technologies.
Nano-Invests.de	An investment portal for nanotechnology promoting an expert-forum, through monthly nano-letter, investor relations/investor marketing for nanotech companies.
Ardesta	Develops industry-building resources such as trade publications, Web sites, and trade shows.

TABLE 6.4 (*Continued*)

Networking and Facilitating Development and Commercialization in the Nanotechnology Sector

MIT-Stanford Venture Laboratory (Vlab)	A volunteer-run, nonprofit organization founded by the MIT Enterprise Forum, the Stanford Office of Technology Licensing, and the Alumni Association of the Graduate School of Business with 20 additional chapters of the MIT Enterprise Forum, Inc. worldwide to foster the growth and success of entrepreneurial ventures by connecting emerging ideas, technology, and people.
Texas Nanotechnology Initiative	A consortium of industry, universities, government, and venture capitalists whose goal is to position Texas as the nanotechnology state.
Albany NanoTech	A global research, development, technology deployment, and education resource supporting accelerated high technology commercialization and job creation through leveraged partnerships between business, government, and academia.
Molecular Manufacturing Shortcut	Promotes developing nanotechnology as a way to facilitate space exploration and settlement.
LARTA	Runs programs throughout the year to help professional investors connect with promising cutting-edge technology companies.
San Francisco Consulting Group	Offers strategic marketing, consulting, and programs for nanotechnology companies and investors.
NVST	Provides entrepreneurs tools and services to showcase their opportunities while providing investors access to the deals they want.
CMP-Cientifica	Provides scientific and technology information through scientific networks of scientists, businesses, and investors active in the technology world and offers strategic consultancy, business intelligence, and investment appraisals.
YASHNANOTECH	A business information and consulting arm of Yash Management and Satellite Ltd. specializing in nanotechnology.
Center for NanoSpace Technologies	A nonprofit research foundation chartered to conceive, establish, and conduct cutting-edge technology, research, and development for infusion into the aerospace, education, energy, life sciences, and shipping/transportation industries.
The Nanotechnology Institute	A collaborative enterprise among academic and research institutions, corporate partners, private investors, government, and economic development. A multi-state initiative involving Pennsylvania, New Jersey, Delaware, and Maryland supported by emerging international alliances with Japan, Italy, and United Kingdom. A multiinstitutional/disciplinary research and development approach to facilitate the transfer and

(Continued)

TABLE 6.4 (*Continued*)

Networking and Facilitating Development and Commercialization in the
Nanotechnology Sector

	commercialization of discoveries and intellectual knowledge that support rapid application of nanotechnology to the life sciences sector and the creation of new enterprises organized around this technology. Lead organizing partners: Ben Franklin Technology Partners of Southeastern Pennsylvania, Commonwealth of Pennsylvania, Drexel University, and the University of Pennsylvania.
Technolytics	A privately held think-tank that investigates technologies that will mature within a two- to three-year period, assesses their impact on business outcomes, develops business strategies that take advantage of the technological changes, and defends against disruptive technologies.
Glocap Tech	An executive search and recruiting firm that places senior engineers, scientists, researchers and management into MEMS, microsystems and nanotechnology companies.
Volant Technologies	Enables disruptive technologies in microsystems, MEMS, and nanotechnology.
The Stanton Group	Specializes in the recruitment and placement of research and development professionals and executives in technology leadership roles especially in nanotechnology and MEMS, materials science, and electrical, mechanical, and software engineering.
iManage Collaboration Software	Provides collaboration software for business, commerce, and supply chain collaboration solutions.
Gaebler Ventures	Business incubator and holding company providing venture capital investment to early-stage companies.
MicroPowder Solutions	A high-technology company with core competencies in nanomaterials, nanotechnology, ultra fine materials, advanced ceramics, and technical due diligence.
TechVision21	Facilitates procurement of grants from U.S. governmental agencies.
Micro and Nanotechnology Commercialization Education Foundation—MANCEF	Globally supports the creation, exchange, and dissemination of knowledge in the commercialization of miniaturization technologies.
Nick Massetti Consulting	Offers technology risk assessment and mitigation for venture capitalists and startups.
Battery Ventures	Broad industry expertise and capital to launch the next generation of category leaders.
Polaris Venture Partners	Backs and plays lead role in developing information and medical technology companies in the U.S. and Europe.
vFinance	Global portal for companies in need of capital and private investors seeking quality deal flow.
Molecular Manufacturing Enterprises	Facilitates accelerated advancements in nanotechnology.

TABLE 6.4 (*Continued*)

Networking and Facilitating Development and Commercialization in the
Nanotechnology Sector

The Nanotechnology Group	A consortium of nano companies, universities, and organizations focusing on development of education solutions in integrated nano sciences.
*n*ABACUS Partners	Provides consultation to governments and industries, to build a portfolio of strategic investments and assist other investors in their identification, management, and exit from nanotechnology ventures.
Newbridge Nanotechnology Index (NYSEArca: NNIX)	Tracks the stock market performance of companies that are currently active in nanotechnology.
Invest Australia	Australian government's inward investment agency acting as the first national point-of-contact for investment inquiries, offering free, comprehensive, and confidential assistance.
CeNTech	Promotes the formation of start-up companies that originate from university research and provides the general conditions for entrepreneurs to further develop their research ideas into marketable products and supports the expansion of companies in the nanotechnology sector.
Cenamps	Promotes commercialization of emerging technologies through collaboration of science with business, through sustainable innovation, and technological development.
NanoBioNet—The Center of Excellence of Nanobiotechnology	Network including partnering universities, research institutes, hospitals, private companies, and experts from the areas of technology transfer and patents as well as from economics and finance.
European Nanotechnology Trade Alliance (ENTA)	Represents the interests of its members across Europe with business interest in nanotechnology.
Cleantech Venture Network	Membership organization bringing insight, opportunities, and relationships to investors, entrepreneurs, and service providers interested in clean technology.
Isha Nanotech	Specializes in growing nanotechnology ventures and in helping investors evaluate nanotechnology opportunities.
SK Helsel & Associates	Accelerating business development by connecting emerging technology companies especially in biotechnology, nanotechnology, optics, and photonics, with decision makers and opinion leaders in business, finance, academia, and media.
MITX Nanotech Exchange (MNE)	A division of the Massachusetts Innovation & Technology Exchange (MITX) featuring a broad range of programs aimed at nanotechnology startups, commercial corporations using nanotechnology, universities, and legal, financial, and consultant service providers.

(Continued)

TABLE 6.4 *(Continued)*

Networking and Facilitating Development and Commercialization in the Nanotechnology Sector

ISE-CCM Nanotechnology (TNY)	Includes companies involved in the science and technology of building electronic circuits and devices from single atoms and molecules.
Atlantic Nano Forum	Provides networking and educational opportunities through meetings and exchange of ideas, and establishes meaningful relationships that will result in the creation of the successful companies.
Preciseley Microtechnology Corp.	A microelectromechanical systems (MEMS) technology consulting company to promote collaboration with established MEMS foundries and research institutions throughout North America and Asia.

Data source: http://www.nanotech-now.com/vc-firms.htm (July 28, 2011)

new and existing drugs, diagnostics, and many other life and non-life science products.

The regional partners of Starpharma are

Australia and New Zealand:

> Industrial Research Limited (Wellington)
>
> Institute of Drug Technology (IDT) Australia, Ltd. (Melbourne)
>
> Monash University (Melbourne)
>
> Peter MacCallum Cancer Institute (Melbourne)
>
> RMIT University (Melbourne)
>
> University of New South Wales, Centre for Entomological Research and Insecticide Technology (Sydney)
>
> Victorian College of Pharmacy, Centre for Drug Candidate Optimisation

Europe

> Institute of Organic Chemistry and Biochemistry (Czech Republic)
>
> REGA Institute (Belgium)

United States

> Cincinnati Children's Hospital Medical Center (Ohio)
>
> Dendritic Nanotechnologies, Inc. (Michigan)
>
> Fox Chase Institute for Cancer (Pennsylvania)
>
> Georgetown University (Maryland)
>
> Institute for Antiviral Research, Utah State University (Utah)
>
> National Institutes of Health (NIH), NCI, and NIAID (Maryland)

U.S. Army Medical Research Institute for Infectious Diseases (Maryland)

University of Alabama (Alabama)

Starpharma's strong patent portfolio has immediate licensing opportunities and is actively seeking strategic commercialization partners and licensees for its technologies at all stages of development. Starpharma also seeks to work with companies to enable the incorporation of its platform technologies into new and existing pharmaceutical products. Partnerships to develop Starpharma's technologies could include Starpharma's expertise for the successful development of dendrimer-based products. Starpharma's established skills and expertise in these areas lower the barrier to entry for commercial partners and licensees in the field of dendrimer nanotechnology.

Products based on Starpharma's dendrimer technology are already on the market for diagnostics and laboratory reagents through license arrangements with partners including Siemens and Merck KGaA. Starpharma has signed separate license agreements with Reckitt Benckiser (formerly SSL International) and Okamoto Industries, Inc. to develop a VivaGel®-coated condom. Reckitt Benckiser manufactures and sells Durex® condoms, the market-leading condom brand worldwide. Okamoto is the market leader for condoms sold in Japan, the world's second largest condom market.

In November 2011, Starpharma announced that its terms for phase III trial for its VivaGel bacterial vaginosis treatment were endorsed by both the European EMA and the U.S. FDA. The company secured funding for the trials through an oversubscribed $32 million institutional placement. Around $16 million of this will be put toward the trials.[13]

The combined strengths derived from the strategic management of Starpharm has enabled it to become a leading company in the business of sexual health (VivaGel), drug delivery for pharmaceuticals, animal health, coatings and inks, crop protection and agrochemicals, diagnostics and research reagents, cosmetics, and water treatment.

6.3.2 Case Study: Nano Terra, Inc.

Nano Terra, Inc. is a privately held nano- and micro-technology development company headquartered in Brighton, Massachusetts. Founded in 2005, Nano Terra's business model is based on establishing long-term collaborations with leading manufacturers and marketers in a wide variety of industries, including electronics, aerospace, energy, industrial products, and consumer goods, as well as government agencies by running co-development programs focusing on a particular product or product advancement to drive scientific innovations through strategic commercial partnerships into the market. Nano-Terra's collaborations encourage adoption of technologies by mature businesses. It has taken in $17.2 million of a planned $23.6 million venture-funding round with Berlow and Whitesides

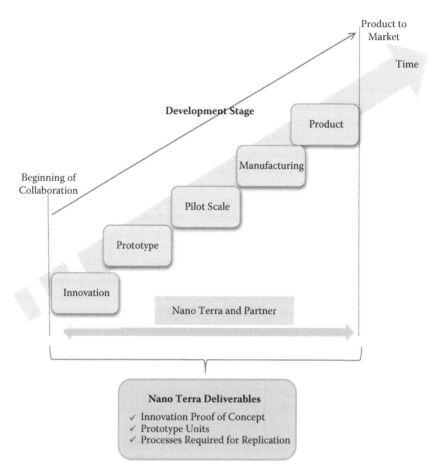

FIGURE 6.1
Nano Terra collaboration model.

investing $3 million. The representation of Nano Terra collaboration model is depicted in Figure 6.1.

Harvard licensed to Nano Terra a broad portfolio of more than 50 issued and pending patents of nano- and micro-scale molecular fabrication methods developed by Professor George Whitesides and the Whitesides Laboratory at Harvard.

The agreement spans the life of the patents. Harvard will receive royalties from products developed from these licensed technologies and will receive equity in Nano Terra. Additional terms of the agreement were not disclosed. The licenses gave Nano Terra exclusive commercialization rights to these technologies in areas outside of the biomedical field such as electronics, aerospace, energy, industrial products, military uses, environmental testing, and consumer goods. The intellectual property licensed from

Harvard—involving molecular self-assembly, rapid prototyping, electrical/optical systems, soft lithography, and microfluidics—has a potentially wide range of applications in a number of important industries.

Key components of Nano Terra's expertise include surface engineering techniques (such as soft lithography, self-assembly, and surface chemistry), and the application of novel nanomaterials. Nano Terra assembles structures and chemistries at all length scales (from nano to micro) on flat, curved, smooth, and uneven surfaces made of a broad range of materials, including metals, polymers, oxides, and ceramics.

The Nano Terra team works with a partner to devise a program to develop breakthrough innovations with an integrated, multidisciplinary team of world-class chemists, biologists, physicists, and engineers. During the program, Nano Terra leads through the innovation and prototyping stages, with the partner taking a more active role as the project moves into the pilot stage. As programs move toward commercialization, where the partners lead the efforts, Nano Terra continues to engage with the partner to build on the teachings from the pilot and initial manufacturing stages. All efforts are dedicated to moving each innovation into the market. The partner pays Nano Terra fees that cover the cost of development (i.e., scientist time, supplies, materials and equipment, etc.). Nano Terra makes no profit on its co-development projects, reducing costs and ensuring more efficient use of the partners' resources and alignment of objectives. Once products are launched, Nano Terra receives revenues from its partners based on the value that the co-development efforts bring to its partners.

Nano Terra's co-development business model enables cost-effective integration of breakthrough technologies into mature businesses. Examples of such partnerships are as follows.

2007: Nano Terra entered into a similar, multiyear development and licensing agreement with 3M Co. of Minnesota, aimed at helping 3M bring to market nano-scale materials and products developed using Nano Terra's molecular fabrication methods.

October 2008: Nano Terra and German drug maker Merck KGaA upgraded their existing product development partnership to a commercialization deal. Nano Terra helped co-develop Merck's "printable electronics" using nanometer-sized materials and a soft lithography technique called micro-contact printing.

March 2009: Nano Terra and Exide Technologies struck an alliance to combine their expertise to develop more efficient and competitive energy storage solutions for network power and transportation solutions. Nano Terra will be bringing its expertise in surface chemistry and surface engineering to Exide's current product line of energy storage solutions.

May 2009: Nano Terra signed a long-term partnership agreement with Honeywell Aerospace to develop functional surfaces for a variety of aerospace applications including the fabrication of thin films for certain military and commercial aerospace applications designed specifically to enhance performance by combining Honeywell's aerospace technology and systems integration expertise with Nano Terra science in functionalization of surfaces.

July 2009: Pentair, Inc. and Nano Terra formed a strategic alliance to jointly develop water-treatment technologies.

April 2011: Nano Terra acquired drug developer Surface Logix with expertise in the Pharmacomer Technology Platform, which enables the discovery of new, highly optimized small molecule drugs, and the rapid identification of preclinical drug candidates. This acquisition opens the door to the use of Nano Terra's technology in a variety of healthcare products where microfabrication and surface chemistry offer tremendous promise in dramatically improving the functionality of healthcare products. It is a logical extension of Nano Terra's business. The combination of Surface Logix's promising therapies and Nano Terra's ability to identify, recruit, and fully partner with major companies will bring life-saving and life-enhancing products more quickly to the people who need them.

April 2011: Kadmon Pharmaceuticals LLC, exploring new avenues in molecular biology to develop therapies that target the metabolomic or signaling pathways associated with disease, including novel anti-hepatitis C therapies, and Nano Terra entered into an agreement under which Kadmon has been granted a perpetual, worldwide exclusive license to three novel, clinical-stage product candidates owned by Nano Terra as well as rights to Nano Terra's drug discovery platform, Pharmacomer™ Technology. The product candidates and technology platform will be transferred to a newly formed joint venture, NT Life Sciences ("NT Life"), co-owned by Kadmon and Nano Terra, which will act as the licensor and recipient of any licensing or royalty fees subject to the agreement. Terms of the agreement were not disclosed. The product candidates and Pharmacomer Technology platform were developed by Surface Logix, Inc., recently acquired by Nano Terra.

This agreement synergistically combines Nano Terra's expertise of its chemistry and nanotechnology with Kadmon's scientific and clinical expertise, unique insights on drug development, as well as a record of significant value creation into the life sciences to open up new opportunities for speedy commercialization to the market.

July 2011: Nano Terra with U.S. Equity Holdings, which has a proven track record in the solar photovoltaic (PV) industry and considerable experience in bringing game-changing products, announced the formation of Microline PV LLC, which will develop metalliza-tion technology to achieve significant performance improvements to crystalline silicon solar photovoltaic cells without capital invest-ment in new equipment and commercialize the PV technology. Microline PV's licensed technology was developed by Nano Terra and this new technology can be used directly in equipment that is already deployed in nearly every present crystalline silicon PV manufacturing facility and is compatible with the most commonly used silver pastes, a key component in the fabrication of PV cells.

While building value for both Nano Terra and its corporate and govern-ment partners, Nano Terra uses its patents to develop and apply technology to address partners' specific business needs. Through this innovative col-laborative business model, Nano Terra enable its partners to gain access to the technology and expertise of Nano Terra for minimal development costs and Nano Terra and its corporate partners share in the value created by the resultant technology.

6.3.3 Case Study: Oxford Instruments plc

Oxford Instruments plc ("Oxford Instruments"), the leading provider of high technology tools and systems for industry and research, has been strengthening its nanotechnology business arm by acquiring complemen-tary businesses. In June 2011, Oxford Instruments acquired Germany-based Omicron NanoTechnology GmbH (together with its subsidiaries in the United Kingdom, United States, Japan, and France; "Omicron") and U.S.-based Omniprobe, Inc. ("Omniprobe").

With a portfolio of 111 patents/patent filings (as of December 31, 2011 in Espacenet worldwide database) Omicron was acquired for a cash consider-ation of €32.4 million (around £28.1 million), on a cash-free, debt-free basis, to include land and buildings valued at €5.0 million (around £4.3 million). Omicorn designs and produces advanced microscopes and chemical analysis instruments for nanotechnology research. Its products are used by scien-tists for research into the properties of materials using ultra high vacuum to ensure extreme sample purity. The microscopes can image and move indi-vidual atoms allowing the observation and manufacture of nanostructures and are used in applications such as research into next generation computer storage and processing devices.

Omicron generated earnings before interest and tax (EBIT) of €3.2 mil-lion in the financial year ended December 31, 2010 from revenue in the same year of €37.9 million. As of December 31, 2010, Omicron had gross assets of €29.3 million.

Omniprobe, with a portfolio of 40 patents/patent applications (as of December 31, 2011 in Espacenet worldwide database), was acquired for an initial cash consideration of US$19.2 million and a deferred element, payable in two-years' time on the terms set out in the agreement, of US$0.8 million (in total around £12.1 million), on a cash-free, debt-free basis. Omniprobe designs and produces tools giving customers nano-scale laboratory capabilities within microscopes. Its products enable probing, manipulation, selective deposition and etching, and sample preparation at the nano-scale for customers in research sectors including nanotechnology and semiconducting devices. A key application is the use of probes to cut out and remove ultra small pieces of a silicon wafer for quality control testing in the semiconductor industry. Omniprobe generated EBIT of $1.7 million in the 12 months to March 31, 2011 from revenue in the same period of $9.1 million.

The combination of Oxford Instruments and these businesses with their patent portfolios strengthens the group's nanotechnology tools sector, through the addition of a complementary portfolio of products, technologies, patent portfolios, and expertise. Both acquired businesses share a complementary skill sets, similar customer base, and routes to market with the existing companies in the group's nanotechnology tools sector. The acquisitions, together with the group's existing capabilities, bring opportunities for the development of integrated new products.[14]

6.3.4 Case Study: Nanofibers—Xanofi

The nanofiber products market was $80.7 million in 2009, according to BCC Research. That market is forecast to reach $2.2 billion in total revenues by 2020. Mechanical and chemical applications account for more than 70% of that market right now. Medical applications will have a significant place in that growing market.

Nanofibers have wide ranging applications in medical applications such as bandages employing nanofibers that consist of different layers that can absorb fluids, deliver antibiotics, and stop blood; surgical mats to protect organs from surgical adhesions; in regenerative medicine to form scaffolding-like structure that gives cells a place to grow; nanofiber patch for the heart that is intended to help the regeneration of dead cardiac tissue, etc.

Xanofi spun off from North Carolina State University, where the technology for nanofibers was developed from seven years of research in the chemical engineering department. The company was founded in September 2010 and raised $300,000 in Angel investment. Xanofi now has a pilot machine at its Raleigh facility. On June 8, 2011, Xanofi's Board of Directors unanimously approved $1.5 million to transition from a proof-of-concept phase into full commercial production. The funds will be used for equipment scale-up, company infrastructure, sales and marketing, ISO certification, and product development. Xanofi has raised money through angel funding in preference to seeking grants from diverse sources.

According to a report published in June 2011, startup Xanofi has developed a new manufacturing technology, XanoShear™, that makes nanofiber production faster, less expensive, and scalable. Xanofi's patent-pending technology can produce nanofibers specifically for medical applications, at half the cost of current manufacturing techniques such as Electrospinning and Meltblowing, with 20 times higher yield. It also has technology for surgical mats (XanoMat™).

Xanofi is currently talking to companies in various industries that may be interested in licensing the technology and that could integrate nanotechnology in medicine.

Although commercialization in medical applications looks promising, Xanofi has strategically decided to first commercialize the technology platform in nonmedical uses such as water filtration, acoustics, and batteries, and is seeking co-development agreements or manufacturing contracts.

The company has recognized that clinical testing is expensive, takes a long time, and is an area in which the company has no expertise and does not want to be entangled in regulatory issues. Medical uses are likely to come from a partner interested in developing the nanofiber manufacturing technology for medical applications.[15]

6.3.5 Case Study: Infineon, Genus, UAlbany Center of Excellence in Nanoelectronics

In January 2004, Infineon, Germany, Genus, United States, and the UAlbany Center of Excellence in Nanoelectronics, United States, entered into a $12 million, three-year partnership to develop nanoscale computer-chip memory devices. Researchers from all three partners would work together at the UAlbany Center of Excellence in Nanoelectronics to optimize atomic layer deposition processes for metal electrode and high-k dielectric materials for sub-45 nm DRAM capacitors on a Genus 300 mm wafer bridge cluster tool.[16]

6.4 Lessons from the Case Studies

These case studies clearly demonstrate the need for an integrated approach involving conceptualization, deriving initial funding, organizing technical and financial support, striking vertical or horizontal strategic partnerships, continually creating, transacting, and managing IPRs, addressing regulatory sensitivities and seeking timely clearances, seeding and nurturing human resources at diverse levels, handling governmental agencies deftly, sensing and acting on the market ecosystem, iteratively amending the business trajectory based on continual learning, and exploiting the evolutionary intelligence for nano-businesses to successfully hatch, grow, survive, and vibrantly sustain the torturous time variant undulated pathways.

6.5 Nanotechnology Patentability Issues: Gray Areas

Article 27 of the TRIPS Agreement defines the minimum standards for Patentable Subject Matter as:

1. Subject to the provisions of paragraphs 2 and 3, patents shall be available for any inventions, whether products or processes, in all fields of technology, provided that they are new, involve an inventive step and are capable of industrial application. Subject to paragraph 4 of Article 65, paragraph 8 of Article 70 and paragraph 3 of this Article, patents shall be available and patent rights enjoyable without discrimination as to the place of invention, the field of technology and whether products are imported or locally produced.

2. Members may exclude from patentability inventions, the prevention within their territory of the commercial exploitation of which is necessary to protect *ordre public* or morality, including to protect human, animal or plant life or health or to avoid serious prejudice to the environment, provided that such exclusion is not made merely because the exploitation is prohibited by their law.

3. Members may also exclude from patentability:

    a. diagnostic, therapeutic, and surgical methods for the treatment of humans or animals;

    b. plants and animals other than microorganisms, and essentially biological processes for the production of plants or animals other than nonbiological and microbiological processes. However, Members shall provide for the protection of plant varieties either by patents or by an effective *sui generis* system or by any combination thereof. The provisions of this subparagraph shall be reviewed four years after the date of entry into force of the WTO Agreement.

One needs to examine the exclusions that are allowable by the TRIPS Agreement vis-à-vis the range of inventions in nanotechnology.

In nanotechnology and especially when a "bottoms up" approach is used in nanobiology, it is possible to interpret some of the processes to be "nano-biological" as opposed to "micro-biological" and hence may be considered to fall within the ambit of the exclusions to patentability.

Inventions related to nanomedicine and regenerative medicine have already started to indicate the impact of nanomaterials on stem cell growth and differentiation, construction of tailored structures to result in targeted tissues, leading to growth and construction of organs that are made of living matter and therefore considered as promising candidates for organ repair or replacement of damaged human organs to function to their natural capabilities or even to function with enhanced capabilities. Such inventions may also be interpreted as falling within the ambit of the exclusions to patentability on the basis of their not complying with "morality" requirements. Developments in nanomedicine exploiting the properties of nanomaterials either as probes/sensors for diagnosis or as agents for directed growth of tissues has opened the doors to unique diagnostics and therapies. Such blurring of boundaries between *in vitro* and *in vivo* processes are raising doubts on whether such methods should be considered as diagnostic, therapeutic, and surgical methods for the treatment of humans or animals as they are being practiced on the human or animal body and therefore may be excluded from patentability. Questions are also being raised on the patentability of nanotechnology methods for diagnostics that are carried out in multiple phases with some steps executed with interaction with humans or animals. Further, the techniques in nanotechnology that lead to regeneration/repair of tissues have applications in cosmetics where again the challenge lies in differentiating whether the repair was directed to treat a disease or was purely cosmetic in effect. Serious debates have been initiated in the use of nano-taggants as sensors and nanobots in humans or animals as they become amenable to being monitored, thereby severely interfering with human rights and impinging on human dignity. Such an interpretation may lead to the inventions being classed as nonpatentable on grounds of morality and in some cases against public order as well.

An area in patenting that is yet to impact nanotechnology patents and their enforcement is the provision of "compulsory licensing." It is likely that several nanotechnology patents may become the subject of compulsory licensing under patent laws of some countries on the grounds that they have been worked and that after three years of their grant, the reasonable requirements of the public with respect to the patented invention have not been satisfied.

Another matter of significance is the impact of nanomaterials on the environment. Regulatory bodies have already raised concerns about the potential adverse environmental, health, and safety (EHS) implications of nanoscale materials.

For example, as an illustration the following could be causes for major regulatory concerns:

- A regulatory pathway could vary with the size of the entity and therefore a nano-sized entity may compete in a neurotransmitter, that is, in normal receptor binding processes, and cause problems.

- New forms of known chemicals such as nanotubes of carbon of a particular size might act like asbestos and therefore cause problems similar to those caused by asbestos.

- Although regular silver is considered inert, nanosilver is biologically very active and therefore may pose an environmental risk in aquatic settings.

Once again, if an invention involves nanomaterials that may seriously prejudice the environment, the invention may be interpreted to fall within the ambit of the exclusions to patentability.

Regulatory authorities around the world are already active in their attempts to set regulation standards for applications involving nanotechnology and a lot is yet to be done in this sensitive area that will determine the course of nano patenting and commercialization.

The issues indicated previously pose challenges to the construction of patent claims that may not fall within the exclusions of patentability. The rapidly developing fields of nanomedicine and nanomaterials pose challenges to the drafting and defense of patent claims for their novelty, inventive step, industrial applicability, and staying off the ambit of the exclusions. The future that lays ahead for patenting in nanotechnology is increasingly becoming complex, murky, and uncertain.

6.6 Institutional IPR Policy and Management

It has been well established that the knowledge-led activities in nanotechnology have led to a significant rise in symbiotic relationships through contract research, sponsored collaborations, and trans-border inter-institutional and intra-institutional cross-functional teams working with multi-institutional collaborative participation in knowledge networking. It is imperative that sustained competitive advantage of nanotechnology-based businesses strongly correlate with the ability of organizations to create, manage, and market "value-added" intellectual assets to derive "first to market" advantage. Further, the role of institutionalized management of IPR based on institutional IPR policy is more than obvious as harmonious interlinked working is nurtured in a formalized framework for "knowledge ownership," knowledge sharing, and "fair benefit sharing" of commercialization between the collaborating partners. The case studies also show that the recipe for a successful nanotechnology business is an intimate mix of factors related to accessing, creating, generating, applying, and trading knowledge with intense IPR protection and enforcement. Early recognition of the knowledge ownership girds to ensure that there are no infringing overlaps and, if any, timely steps are initiated to avoid litigations and ensure facile move of innovations to the market.

The aim of any institutional IPR policy is to create a facilitating ecosystem with transparent guidelines and benchmarks for ownership, protection, and commercialization of the developed IP while at the same time upholding the core moral values such as integrity, merit, researchers' freedom, and excellence. Such an institutional IPR policy should provide transparent guidelines at least on

- business mission, objectives, and the role of IPR in business;
- institutional body responsible for the management of IPR;
- scope and applicability of the IPR policy;
- what constitutes conflict of interest;
- ownerships and assignment of rights;
- confidentiality and disclosure of privileged information;
- invention disclosure to the institution and documentation;
- publication policy and communications policy;
- information security in digital and nondigital media;
- prior art search, review, and evaluation of innovations for their IPR protection including strategizing approaches to avoiding statutory exclusions to patentability (if invention in principle is assessed to be patentable by the internal team of experts), the generation of "freedom to operate" reports for the developed innovations so as to not knowingly infringe on others' IPR;
- infrastructure and process for seeking IPR for the developments including criteria for foreign filings, procedures to be followed, utilization of international IPR conventions, etc.;
- handling proceedings in patent offices for prosecution of patent applications, oppositions, etc.;
- minimum expectations from due diligence on others' IPR;
- contracts and agreements for interinstitutional dealings and IPR transactions including jurisdiction;
- commercialization of innovations and IPR;
- principles of benefit sharing;
- dispute resolution;
- handling of legal issues as infringements, damages, liability, and indemnity insurance;
- IPR audit;
- consequences for compliance of institutional policy.

Thus, the management of intellectual property rights based on the institutional IPR policy has to be designed to weave the business mission and objectives with the innovation process stringed from concept to development, commercialization, and postcommercialization phases to extract the maximal value of the investments. A few representations (Figure 6.2 to Figure 6.5) illustrate the key features that are of significance in any IPR management process.

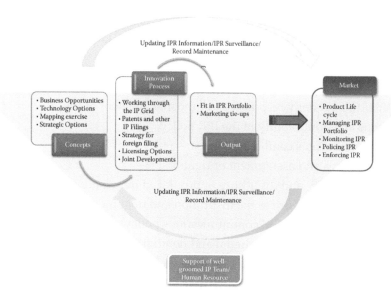

FIGURE 6.2
Managing intellectual property.

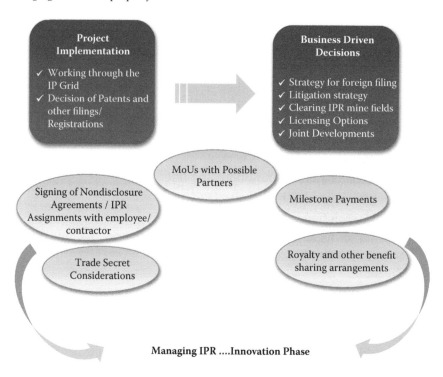

FIGURE 6.3
Managing intellectual property—Innovation Phase.

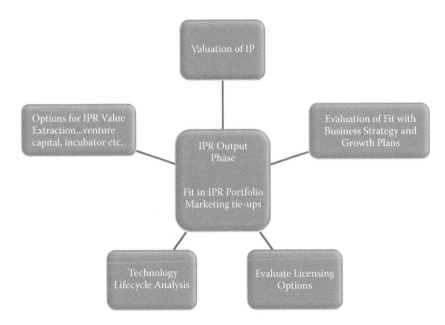

FIGURE 6.4
Managing intellectual property—Output Phase.

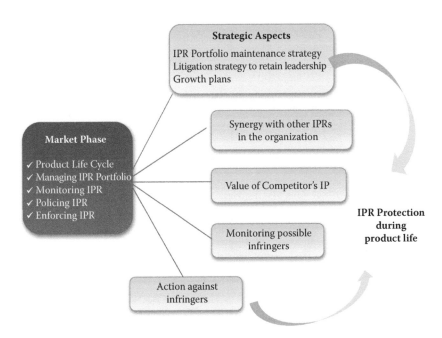

FIGURE 6.5
Managing intellectual property—Market Phase.

6.7 IPRinternalise®

A model "IPRinternalise®" seamlessly integrates IPR in the innovation process in a well-structured process providing an experience-led framework with value-added learning during the process of innovation. "IPRinternalise" starts at the problem definition phase where a knowledge seeker is induced to explore and ethically exploit with appropriate acknowledgment the richness of existing knowledge (prior art), contextually build on it, and provide technical solutions to the identified problem as he or she assesses it, and in the process inculcates the necessary ethical platform and skills to assess the existing prior art in all forms of literature including patent literature and then proceeds to create solutions and subsequently protect the creation. This model is depicted in Figure 6.6.

"IPRinternalise®" also provides the knowledge seeker an opportunity to conduct an early stage assessment of his innovations and a check against possible infringement of others' IPR, and design his innovation path to ensure that his solutions with built-in ethics and respect for others' IPRs have the much needed "freedom to operate" (FTO) in the geographies of his interest. "IPRinternalise" provides a problem-based approach to IP, which for a knowledge seeker is "stress and burden free" but "relevant and need based," as one is drawn into it by a natural tide of the innovation process originating from one's immediate requirements and ethically addressing the implications of one's innovations in the later stages of the innovation cycle within a moral framework of societal acceptance.

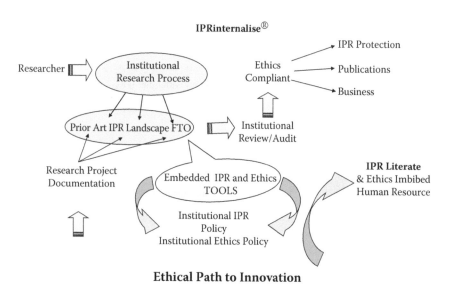

Ethical Path to Innovation

FIGURE 6.6
Model IPRinternalise®.

6.8 Nanotechnology: Sunrise of the Day After

Every sunrise in the last few decades has been a boost to nanotechnology that has brought several innovations to light. The next decade has been predicted to be of even higher luminosity although checkered with a few shadows of IP fights and tribulations. Lessons have been learned from the past, but steps ahead may require transformations through directed unlearning and partial dissociation from the past to enable the next quantum "nano" jump when the sun rises the day after. Laws will have to be crafted to support such laudable efforts in an intensely activated human-molecule interactive nanoworld. The new rules of the market ecosystem will demand creative and cooperative competition that will see higher levels of precompetitive collaboration operating with higher levels of cautious openness to collectively maximize delivery from the minimal resources.

References

1. http://uts.cc.utexas.edu/~bramblet/ant301/four.html
2. Evident clears bankruptcy court; $75K/month fed grant in works, April 12, 2010, http://www.bizjournals.com/albany/stories/2010/04/12/story6.html
3. Luna Innovations emerges from Chapter 11 reorganization, January 12, 2010, http://ir.lunainnovations.com/phoenix.zhtml?c=196907&p=irol-newsArticle2&ID=1374345&highlight=
4. Ratner, D., It's a small word after all, *USA Today*, May 2004, http://findarticles.com/p/articles/mi_m1272/is_2708_132/ai_n6021788/
5. Yu, S., Navigating the nanotechnology patent thicket, Medical Device and Diagnostic Industry, November 2007, http://www.mddionline.com/article/navigating-nanotechnology-patent-thicket
6. Crothers, B., Intel to pay Nvidia $1.5 billion in licensing fees, *CNET News*, January 10, 2011, http://news.cnet.com/8301-13924_3-20028066-64.html
7. Lee, A., Examining the viability of patent pools to the growing nanotechnology patent thicket, master's thesis, University of Virginia, http://www.nanotech-project.org/process/assets/files/2722/70_nano_patent_pools.pdf.
8. Marx, V., Poised to branch out, *Nature Biotechnology*, 26, 729–732, 2008, http://www.nature.com/nbt/journal/v26/n7/box/nbt0708-729_BX1.html
9. Rasmussen, B., *Innovation and Commercialisation in the Biopharmaceutical Industry, Creating and Capturing Value*, Edward Elgar Publishing Limited, Cheltenham, U.K., 2010.
10. Starpharma acquires Dendritic Nanotechnologies, *Genetic Engineering & Biotechnology News*, October 4, 2006, http://www.genengnews.com/gen-news-highlights/starpharma-acquires-dendritic-nanotechnologies/6500802/
11. Starpharma and Industrial Research Ltd. establish joint venture, March 2005, http://www.glycosyn.com/news/starpharma-and-industrial-research-ltd-establish-joint-venture

12. GSK awarded funds to advance dermal treatment with Starpharma's dendrimers, August 22, 2011, http://www.solubility.com.au/images/110822_GSKStarpharmaDendrimers.pdf

13. Bushell-Embling, D., Starpharma closer to VivaGel approval in Europe and US, *Life Scientist*, November 23, 2011, http://www.lifescientist.com.au/article/408325/starpharma_closer_vivagel_approval_europe_us/#comments

14. Oxford Instruments acquires Omicron NanoTechnology GmbH and Omniprobe, Inc., June 14, 2011, http://www.oxinst.com/investors/Additional%20Documents/Oxford-Instruments-acquisitions-announcement-june-2011.pdf

15. Vinluan, F., A nanofiber company's solution could advance nanotechnology in medicine, *MEDCITY News*, June 21, 2011, http://www.medcitynews.com/2011/06/a-nanofiber-companys-solution-could-advance-nanotechnology-in-medicine/

16. Technology update, January 30, 2004, http://nanotechweb.org/cws/article/tech/18946

Index

A

Abbott Laboratories, 230
ACM Technologies, Inc., 174
AcryMed, Inc., 86, 87
Advanced Medical Solutions Limited, 86
Advent Technologies, 132
Agrochemical program, 233
AIM, *see* Alternative Investment Market
Albany NanoTech, 235
Alternative Investment Market (AIM), 222, 223
Angel investors, 123
Angiotech, 36
Angstrom Partners, 130
Antimicrobial paint, 44–45
Apple, 227
Applied Materials Ventures, 234
Applied Nanotech Holdings, Inc., 19, 49, 50, 100–108
Arcturus Capital, 131
Ardesta, 234
Arrowhead Research Corporation, 131
Asian Institute of Technology, 86
AS Russian Academy of Science Institute of Theoretical and Applied Mechanics, 87
Atlantic Nano Forum, 238
Atomistix, 218
Austin Ventures, 131
Australian Patents Act, 36
Axiom Capital Management, 130

B

Barter rights, 93
BASF v. Orica Australia, 31
Battelle Memorial Institute, 20
Batteries, nano-enabling, 7
Battery Ventures, 236
Bayer, 228
Benchmarks

capability, 38
ownership, 248
patentability, 25
Bilcare Research, Pune, India, 111–113
Bilcare Technology, 113
Biogen Inc. v. Medeva plc, 35
Bio Life Technical, 234
BioMers Pte Ltd., 125–126
Black holes of knowledge, 1
Bruker Corporation, 165
Business perspectives, patent-led, 91–134
    angel investors, 123
    automotive catalytic converters, 127
    barter rights, 93
    bonds, 124–134
        BioMers Pte Ltd., 125–126
        Holmenkol AG, 126
        ItN Nanovation AG, 126–127
        Lumiphore, Inc., 128
        MagForce AG, 125
        Membrane Instruments Technology Pte Ltd., 128–129
        Microlight Sensors Pte Ltd., 128
        Namos GmbH, 127–128
        Nanosys, Inc., 129–134
        venture capital providers and angels, 130–134
    case studies
        Applied Nanotech Holdings, Inc., 100–108
        Bilcare Research, Pune, India, 111–113
        HyCa Technologies Pvt Ltd., Mumbai, India, 113–114
        Innovalight and DuPont merger, 118–120
        mPhase Technologies, Inc., 109–111
        NanoInk Inc., 96
        Nanosphere, Inc., 96–97
        Nanostart AG, 123–124
        NVE Corporation, 97–99

255

Cooperative research and development
    agreement (CRADA), 109
Copy Technologies, Inc., 174
CRADA, *see* Cooperative research and
    development agreement
Cross licensing, 226
C sixty, 95
CW Group, 131

D

DARPA, *see* Defense Advanced
    Research Projects Agency
Database(s), 72
    assignment of roles, 82
    commonly used, 71
    Espacenet worldwide, 243
    inception of, 6
    INDEX feature and, 83
    paid, 88
    purpose, 67
Database, Science and Technology
    Network, 81
Defense Advanced Research Projects
    Agency (DARPA), 216
Dendrimer, 228
Dendritic NanoTechnologies, Inc., 231
Dip Pen Nanolithography® (DPN®), 96
Direct Billing International, Inc., 174
DNA sequencing machines, maker of,
    122
Dow Chemical, 228
DPN®, *see* Dip Pen Nanolithography®, 96
DSM, 228
DSM v. 3D Systems, 158–162
Due diligence process, 123
DuPont, 95, 118–120
*DuPont Air Products NanoMaterials
    LLC v. Cabot Microelectronics
    Corporation*, 137–142
Dupont Somos, 159
D-Wave Systems Inc., 19
Dynamic nanovation, 5, 6

E

EAPC, *see* Eurasian Patent Convention
Earnings before interest and tax (EBIT),
    243

Eastman Kodak Company, 20
EBIT, *see* Earnings before interest and
    tax
ECLA, *see* European Classification
    System
EHS implications, *see* Environmental,
    health, and safety
    implications
E. I. Du Pont De Nemours and
    Company, 19
EIS Office Solutions, Inc., 174
*Elan Pharmaceuticals International Ltd.
    v. Abraxis BioScience, Inc.*,
    146–151
Electronic theses and dissertation
    (ETD), 72
Electrospinning, 245
Electrowetting, 109
EMD Biosciences, 232
Energy Innovative Products, Inc., 110
Energy storage, 15
Enterprise knowledge networking, 2
Environmental, health, and safety (EHS)
    implications, 247
EPO, *see* European Patent Office
ETD, *see* Electronic theses and
    dissertation
EU, *see* European Union
Eurasian Patent Convention (EAPC), 55
European Classification System (ECLA),
    68
European Nanotechnology Trade
    Alliance, 237
European Patent Office (EPO), 68–69
    approach to assessing novelty, 28
    argument rejected by, 31–32
    Asylum Research, 164
    Boards of Appeal, 33
    Technical Board of Appeals, 28
European patent system, 57
European Union (EU), 3
Evergreen Solar, 219
EV Group v. 3M, 167–169
Evident Technologies, Inc., 216, 217
*Evident Technologies, Inc. v. Everstar
    Merchandise Company Ltd.*, 135
Evolution Capital, 131
Exxon Mobil, 95